Beginning Apache Cassandra Development

Vivek Mishra

Apress®

Beginning Apache Cassandra Development

ISBN-13 (pbk): 978-1-4842-0143-5

ISBN-13 (electronic): 978-1-4842-0142-8

Managing Director: Welmoed Spahr
Lead Editor: Jeff Olson
Technical Reviewer: Brian O'Neill
Editorial Board: Steve Anglin, Mark Beckner, Gary Cornell, Louise Corrigan, Jim DeWolf, Jonathan Gennick, Robert Hutchinson, Michelle Lowman, James Markham, Matthew Moodie, Jeff Olson, Jeffrey Pepper, Douglas Pundick, Ben Renow-Clarke, Gwenan Spearing, Matt Wade, Steve Weiss
Coordinating Editor: Melissa Maldonado
Copy Editor: Lori Cavanaugh
Compositor: SPi Global
Indexer: SPi Global
Artist: SPi Global
Cover Designer: Anna Ishchenko

Distributed to the book trade worldwide by Springer Science+Business Media New York, 233 Spring Street, 6th Floor, New York, NY 10013. Phone 1-800-SPRINGER, fax (201) 348-4505, e-mail orders-ny@springer-sbm.com, or visit www.springeronline.com. Apress Media, LLC is a California LLC and the sole member (owner) is Springer Science + Business Media Finance Inc (SSBM Finance Inc). SSBM Finance Inc is a Delaware corporation.

For information on translations, please e-mail rights@apress.com, or visit www.apress.com.

Apress and friends of ED books may be purchased in bulk for academic, corporate, or promotional use. eBook versions and licenses are also available for most titles. For more information, reference our Special Bulk Sales–eBook Licensing web page at www.apress.com/bulk-sales.

Any source code or other supplementary materials referenced by the author in this text is available to readers at www.apress.com. For detailed information about how to locate your book's source code, go to www.apress.com/source-code/.

For my wife, Rashmi, and my angel, Uditi.
Without you, none of this would be possible.

Contents at a Glance

Contents

About the Author

Vivek Mishra his a technology enthusiast, author, speaker, and big data architect. He is a lead commitor to the open source project Kundera and has contributed to Apache Cassandra.

Presently, he works as Lead Software Engineer with Impetus Infotech pvt Ltd. He is focused on real-time data quality and virtualization using machine learning algorithms and their applications to address business problems.

About the Technical Reviewer

Brian O'Neill is a technology leader and recognized authority on big data. He leads and contributes to open-source projects involving distributed storage, real-time computation, and analytics. He won InfoWorld's Technology Leadership award in 2013 and was selected as a Datastax Cassandra MVP 2012-2014. He authored the Dzone reference card on Cassandra and recently published a book on distributed processing titled, Storm Blueprints: Patterns for Distributed Real-time Computation. He holds patents in artificial intelligence and data management and he is an alumnus of Brown University.

Presently, Brian is CTO for Health Market Science (HMS), where he heads the development of a big data platform focused on analytics and data management for the healthcare space. The platform is powered by Storm and Cassandra and delivers real-time data management and analytics against thousands of disparate data feeds, including medical claims and social media.

Acknowledgments

First I would like to thanks my wife Rashmi and my daughter Uditi for supporting me throughout my career and all the encouragement to write this book. Special thanks to my parents for encouraging me to choose the right career path and teaching me dedication towards the work. I also want to thank all my colleagues, seniors, and mentors for all the support and thoughtful discussions.

A BIG thanks to Chris Nelson, Brian O'Neill, Melissa Maldonado, and the Apress team for a thorough technical review and amazing experience as an author. Without this team, it wouldn't have been possible.

Introduction

Big or large data has been the talk of the town in recent years. With possibilities for solving unstructured and semi-structured data issues, more and more organizations are gradually moving toward big data powered solutions. This essentially gives organization a way to think "beyond RDBMS." This book will walk you through many such use cases during the journey.

Many NoSQL databases have been developed over the last 4-5 years. Recent research shows there are now more than 150 different NoSQL databases. This raises questions about why to adopt a specific database. For example, is it scalable, under active development, and most importantly accepted by the community and organizations? It is in light of these questions that Apache Cassandra comes out as a winner and indicates why it is one of the most popular NoSQL databases currently in use.

Apache Cassandra is a columnar distributed database that takes database application development forward from the point at which we encounter the limitations of traditional RDBMSs in terms of performance and scalability. A few things that restrict traditional RDBMSs are that they require predefined schemas, the ability to scale up to hundreds of data nodes, and the amount of work involved with data administration and monitoring. We will discuss these restrictions and how to address these with Apache Cassandra.

Beginning Apache Cassandra Development introduces you to Apache Cassandra, including the answers to the questions mentioned above, and provides a detailed overview and explanation of its feature set.

Beginning with Cassandra basics, this book will walk you through the following topics and more:

- Data modeling

- Cluster deployment, logging, and monitoring

- Performance tuning

- Batch processing via MapReduce

- Hive and Pig integration

- Working on graph-based solutions

- Open source tools for Cassandra and related utilities

The book is intended for database administrators, big data developers, students, big data solution architects, and technology decision makers who are planning to use or are already using Apache Cassandra.

Many of the features and concepts covered in this book are approached through hands on recipes that show how things are done. In addition to those step-by-step guides, the source code for the examples is available as a download from the book's Apress product page (www.apress.com/9781484201435).

CHAPTER 1

■ ■ ■

NoSQL: Cassandra Basics

The purpose of this chapter is to discuss NoSQL, let users dive into NoSQL elements, and then introduce big data problems, distributed database concepts, and finally Cassandra concepts. Topics covered in this chapter are:

- NoSQL introduction
- CAP theorem
- Data distribution concepts
- Big data problems
- Cassandra configurations
- Cassandra storage architecture
- Setup and installation
- Logging with Cassandra

The intent of the detailed introductory chapter is to dive deep into the NoSQL ecosystem by discussing problems and solutions, such as distributed programming concepts, which can help in solving scalability, availability, and other data-related problems.

This chapter will introduce the reader to Cassandra and discuss Cassandra's storage architecture, various other configurations, and the Cassandra cluster setup over local and AWS boxes.

Introducing NoSQL

Big data's existence can be traced back to the mid 1990s. However, the actual shift began in the early 2000s. The evolution of the Internet and mobile technology opened many doors for more people to participate and share data globally. This resulted in massive data production, in various formats, flowing across the globe. A wider distributed network resulted in incremental data growth. Due to this massive data generation, there is a major shift in application development and many new domain business possibilities have emerged, like:

- Social trending
- OLAP and Data mining
- Sentiment analysis
- Behavior targeting
- Real-time data analysis

With high data growth into peta/zeta bytes, challenges like scalability and managing data structure would be very difficult with traditional relational databases. Here big data and NoSQL technologies are considered an alternative to building solutions. In today's scenario, existing business domains are also exploring the possibilities of new functional aspects and handling massive data growth simultaneously.

NoSQL Ecosystem

NoSQL, often called "Not Only SQL," implies thinking beyond traditional SQL in a distributed way. There are more than 150 NoSQL databases available today. The following are a few popular databases:

- Columnar databases, such as Cassandra & HBase

- Document based storage like MongoDB & Couchbase

- Graph based access like Neo4J & Titan Graph DB

- Simple key-value store like Redis & Couch DB

With so many options and categories, the most important question is, what, how, and why to choose! Each NoSQL database category is meant to deal with a specific set of problems. **Specific technology for specific requirement** paradigm is leading the current era of technology. It is certain that a single database for all business needs is clearly not a solution, and that's where the need for NoSQL databases arises. The best way to adopt databases is to understand the requirements first. If the application is polyglot in nature, then you may need to choose more than one database from the available options. In the next section, we will discuss a few points that describe why Cassandra could be an answer to your big data problem.

CAP Theorem

CAP theorem, which was introduced in early 2000 by Eric Brewer, states that no database can offer **Consistency, Availability,** and **Partition tolerance** together (see Figure 1-1), but depending on use case may allow for any two of them.

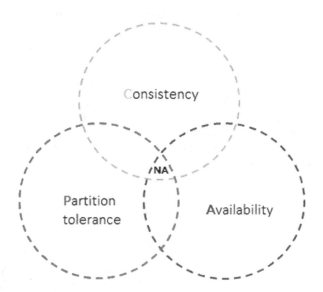

Figure 1-1. *CAP theorem excludes the possibility of a database with all three characteristics (the "NA" area)*

Traditional relational database management systems (RDBMS) provide atomicity, consistency, isolation, and durability (ACID) semantics and advocate for strong consistency. That's where most of NoSQL databases differ and strongly advocate for partition tolerance and high availability with eventual consistency.

> **High availability** of data means data must be available with minimal latency. For distributed databases where data is distributed across multiple nodes, one way to achieve high availability is to replicate it across multiple nodes. Like most of NoSQL databases, Cassandra also provides high availability.

> **Partition tolerance** implies if a node or couple of nodes is down, the system would still be able to serve read/write requests. In scalable systems, built to deal with a massive volume of data (in peta bytes) it is highly likely that situations may occur often. Hence, such systems have to be partition tolerant. Cassandra's storage architecture enables this as well.

> **Consistency** means consistent across distributed nodes. Strong **consistency** refers to most updated or consistent data on each node in a cluster. On each read/write request most stable rows can be read or written to by introducing latency (downside of NoSQL) on each read and write request, ensuring synchronized data on all the replicas. Cassandra offers eventual consistency, and levels of configuration consistency for each read/write request. We will discuss various consistency level options in detail in the coming chapters.

Budding Schema

Structured or fixed schema defines the number of columns and data types before implementation. Any alteration to schema like adding column(s) would require a migration plan across the schema. For semistructured or unstructured data formats where number of columns and data types may vary across multiple rows, static schema doesn't fit very well. That's where budding or dynamic schema is best fit for semistructured or unstructured data.

Figure 1-2 presents four records containing twitter-like data for a particular user id. Here, the user id **imvivek** consists of three columns "tweet body", "followers", and "retweeted by". But on the row for user "apress_team" there is only the column followers. For unstructured schema such as server logs, the number of fields may vary from row to row. This requires the addition of columns "on the fly" a strong requirement for NoSQL databases. Traditional RDBMS can handle such data set in a static way, but unlike Cassandra RDBMS cannot scale to have up to a million columns per row in each partition. With predefined models in the RDBMS world, handling frequent schema changes is certainly not a workable option. Imagine if we attempt to support dynamic columns we may end up having many null columns! Having default null values for multiple columns per row is certainly not desirable. With Cassandra we can have as many columns as we want (up to 2 billion)! Also another possible option is to define datatype for column names (comparator) which is not possible with RDBMS (to have a column name of type integer).

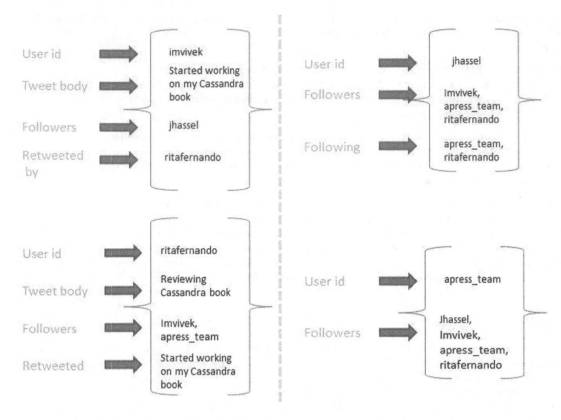

Figure 1-2. A dynamic column, a.k.a. budding schema, is one way to relax static schema constraint of RDBMS world

Scalability

Traditional RDBMSs offer vertical scalability, that is, scaling by adding more processors or RAM to a single unit. Whereas, NoSQL databases offer horizontal scalability, and add more nodes. Mostly NoSQL databases are schemaless and can perform well over **commodity** servers. Adding nodes to an existing RDBMS cluster is a cumbersome process and relatively expensive whereas it is relatively easy to add data nodes with a NoSQL database, such as Cassandra. We will discuss adding nodes to Cassandra in coming chapters.

No Single Point of Failure

With centralized databases or master/slave architectures, where database resources or a master are available on a single machine, database services come to a complete halt if the master node goes down. Such database architectures are discouraged where high availability of data is a priority. NoSQL distributed databases generally prefer multiple master/slave configuration or peer-to-peer architecture to avoid a single point of failure. Cassandra delivers peer-to-peer architecture where each Cassandra node would have an identical configuration. We will discuss this at length in the coming chapters.

Figure 1-3a depicts a system single master acting as single point of contact to retrieve data from slave nodes. If the master goes down, it would bring the whole system to a halt until the master node is reinstated. But with multiple master configurations, like the one in Figure 1-3b, a single point of failure does not interrupt service.

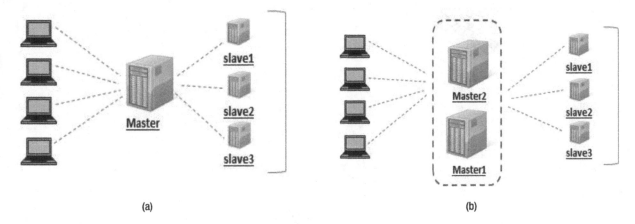

(a)   (b)

Figure 1-3. *Centralized vs. distributed architecural setup*

High Availability

High availability clusters suggest the database is available with 24x7 support with minimal (or no) downtime. In such clusters, data is replicated across multiple nodes, in case one node is down still another node is available to serve the read/write requests until that node is up and running. Cassandra's peer-to-peer architecture ensures high availability of data with co-location.

Identifying the Big Data Problem

Recently, it has been observed that developers are opting for NoSQL databases as an alternative to RDBMS. However, I recommend that you perform an in-depth analysis before deciding on NoSQL technologies. Traditional RDBMS does offer lots of features which are absent in most of NoSQL databases. A couple of questions that must be analyzed and answered before jumping to a NoSQL based approach include

- Is it really a big data problem?
- Why/where RDBMS fails to deliver?

Identifying a "big data problem" is an interesting errand. Scalability, nature of data (structured, unstructured, or semistructured) and cost of maintaining data volume are a few important factors. In most cases, managing secured and structured data within an RDBMS may still be the preferred approach; however, if the nature of the data is semistructured, less vulnerable, and scalability is preferred over traditional RDBMS features (e.g., joins, materialized view, and so forth), it qualifies as a big data use case. Here data security means the authentication and authorization mechanism. Although Cassandra offers decent support for authentication and authorization but RDBMS fairs well in comparison with most of NoSQL databases.

Figure 1-4 shows a scenario in which a cable/satellite operator system is collecting audio/video transmission logs (on daily basis) of around 3 GB/day per connection. A "viewer transmission analytic system" can be developed using a big data tech stack to perform "near real time" and "large data" analytics over the streaming logs. Also the nature of data logs is uncertain and may vary from user to user. Generating monthly/yearly analytic reports would require dealing with petabytes of data, and NoSQL's scalability is definitely a preference over that of RDBMS.

Figure 1-4. *Family watching satellite transmitted programs*

Consider an example in which a viewer transmission analytic system is capturing random logs for each transmitted program and watched or watching users. The first question we need to ask is, is it really a big data problem? Yes, here we are talking about logs; imagine in a country like India the user base is huge as are the logs captured 24x7! Also, the nature of transmitted logs may be random, meaning the structure is not fixed! It can be semi-structured or totally unstructured. That's where RDBMS will fail to deliver because of budding schema and scalability problems (see previous section).

To summarize, build a NoSQL based solution if:

- Data format is semi/unstructured

- RDBMS reaches the storage limit and cannot scale further

- RDBMS specific features like relations, indexes can be sacrificed against denormalized but distributed data

- Data redundancy is not an issue and a read-before-write approach can be applied

In the next section, we will discuss how Cassandra can be a best fit to address such technical and functional challenges.

Introducing Cassandra

Cassandra is an open-source column, family-oriented database. Originally developed at Facebook, it has been an Apache TLP since 2009. Cassandra comes with many important features; some are listed below:

- Distributed database\

- Peer to Peer architecture

- Configurable consistency

- CQL (Cassandra Query Language)

Distributed Databases

Cassandra is a global distributed database. Cassandra supports features like replication and partitioning. Replication is a process where system maintains n* number of replicas on various data sites. Such data sites are called nodes in Cassandra. Data Partitioning is a scheme, where data may be distributed across multiple nodes. Partitioning is usually for managing high availability/performance on data.

■ **Note** A node is a physical location where data resides.

Peer-to-Peer Design

Cassandra storage architecture is peer-to-peer. Each node in a cluster is assigned the same role, making it a decentralized database. Each node is independent of the other but interconnected. Nodes in a network are capable of serving read/write database requests, so at a given point even if a node goes down, subsequent read/write requests will be served from other nodes in the network, hence there is no SPOF (Single Point Of Failure).

Figure 1-5 is a graphical representation of peer-to-peer (P2P) architecture.

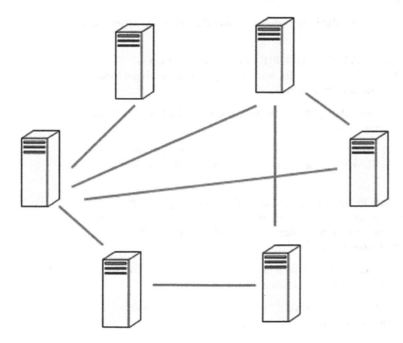

Figure 1-5. *Peer to Peer decentralized Cassandra nodes. Every node is identical and can communicate with other nodes*

Configurable Data Consistency

Data consistency is synchronization of data across multiple replica nodes. Eventually the consistency-based data model returns the last updated record. Such a data model is widely supported by many distributed databases. Cassandra also offers configurable eventual consistency.

Write Consistency

If the data is successfully written and synchronized on replica nodes before acknowledging the write request, data is considered write consistent. However, various consistency level values are possible while submitting a write request. Available consistency levels are

- **ANY:** A write must be written to at least ANY one node. In this case, all replica nodes are down and **"hinted_handoff_enabled: true"** (default is true), then still corresponding write data and hint will be stored by coordinator node, and later once all replica nodes are up, they will be coordinated to at least one node. That written data will not be available for reads until all replica nodes are down. Though ANY is the lowest consistency level but with highest availability as it requires data to be replicated on any one node before sending write acknowledgment.

- **ONE:** With consistency level ONE; write request must be successfully written on at least one replica node before acknowledgment.

- **QUORUM*:** With the consistency level QUORUM* write requests must be successfully written on a selected group of replica nodes.

- **LOCAL_QUORUM:** With the consistency level LOCAL_QUORUM write requests must be successfully written on a selected group of replica nodes, known as quorum, which are locally available on the same data center as the coordinator node.

- **EACH_QUORUM:** With the consistency level EACH_QUORUM write requests must be successfully written on select groups of replica nodes (quorum).

- **ALL:** With the consistency level ALL write requests must be written to the commit log and memory table on all replica nodes in the cluster for that row key to ensure the highest consistency level.

- **SERIAL:** Linearizable consistency is being introduced in Cassandra 2.0 as a lightweight transaction support. With the consistency level SERIAL write requests must be written to the commit log and memory table on quorum replica nodes conditionally. Here conditionally means either guaranteed write on all nodes or none.

- **TWO:** Similar to ONE except with the consistency level TWO write requests must be written to the commit log and memory table on minimum two replica nodes.

- **THREE:** Similar to TWO except with the consistency level TWO write requests must be written to the commit log and memory table on a minimum of three replica nodes.

Read Consistency

No data is of much use if it is not consistent. Large or small data applications would prefer not to have dirty reads or inconsistent data. A dirty read is a scenario where a transaction may end up in reading uncommitted data from another thread. Although dirty reads are more RDBMS specific, with Cassandra there is a possibility for inconsistent data if the responsible node is down and the latest data is not replicated on each replica node. In such cases, the application may prefer to have strong consistency at the read level. With Cassandra's tunable consistency, it is possible to have configurable consistency per read request. Possible options are

- **ONE:** With the read consistency level ONE, data is returned from the nearest replica node to coordinator node. Cassandra relies on snitch configuration to determine the nearest possible replica node. Since a response is required to be returned from the closest replica node, ONE is the lowest consistency level.

- **QUORUM:** With the read consistency level QUORUM, the last updated data (based on timestamp) is returned among data responses received by a quorum of replica nodes.

- **LOCAL_QUORUM:** With the read consistency level LOCAL_QUORUM, the last updated data (based on timestamp) is returned among the data response received by a local quorum of replica nodes.

- **EACH_QUORUM:** With the read consistency level EACH_QUORUM, the last updated data (based on timestamp) is returned among the data response received by each quorum of replica nodes.

- **ALL:** With the read consistency level ALL, the last updated data (based on timestamp) returned among the data response received from all replica nodes. Since responses with the latest timestamp are returned among all replica nodes, ALL is the highest consistency level.

- **SERIAL:** With the read consistency level SERIAL, it would return the latest set of columns committed or in progress. Uncommitted transactions discovered during read would result in implicit commit of running transactions and return to the latest column values.

- **TWO:** With the read consistency level TWO, the latest column values will be returned from the two closest replica nodes.

- **THREE:** With the read consistency level THREE, the latest column values will be returned from three of the closest replica nodes.

Based on the above-mentioned consistency level configurations, the user can always configure each read/write request with a desired consistency level. For example, to ensure the lowest write consistency but the highest read consistency, we can opt for ANY as write consistency and ALL for read consistency level.

Cassandra Query Language (CQL)

One of the key features of Cassandra from an end user perspective is ease-of-use rather than familiarity. Cassandra query language (CQL) was introduced with Cassandra 0.8 release with the intention of having a RDBMS style structured query language (SQL). Since its inception CQL has gone through many changes. Many new features have been introduced in later releases along with lots of performance-related enhancement work. CQL adds a flavor of known data definition language (ddl) and data manipulation language (dml) statements.

During the course of this book, we will be covering most of the CQL features.

Installing Cassandra

Installing Cassandra is fairly easy. In this section we will cover how to set up a Cassandra tarball (.tar file) installation over Windows and Linux box.

1. Create a folder to download Cassandra tarball, for example:

    - Run `mkdir /home/apress/Cassandra` {Here apress is user.name environment variable}

    - Run `cd/home/apress/cassandra`

2. Download the Cassandra tarball:

- Linux: `wget http://archive.apache.org/dist/cassandra/2.0.6/apache-cassandra-2.0.6-bin.tar.gz`

- Windows: `http://archive.apache.org/dist/cassandra/2.0.6/apache-cassandra-2.0.6-bin.tar.gz`

3. Extract the downloaded tar file using the appropriate method for your platform:

- For Linux, use the following command: `tar- xvf apache-cassandra-2.0.6-bin.tar.gz`

- For Windows, you may use tools like WinZip or 7zip to extract the tarball.

■ **Note** If you get an "Out of memory" or segmentation fault, check for the `JAVA_HOME` and `JVM_OPTS` parameters in `cassandra-env.sh` file.

Logging in Cassandra

While running an application in development or production mode, we might need to look into server logs in certain circumstances, such as:

- Performance issues

- Operation support

- Debug application vulnerability

Default server logging settings are defined within the `log4j-server.properties` file, as shown in the following.

```
# output messages into a rolling log file as well as stdout
log4j.rootLogger=INFO,stdout,R
# stdout
log4j.appender.stdout=org.apache.log4j.ConsoleAppender
log4j.appender.stdout.layout=org.apache.log4j.PatternLayout
log4j.appender.stdout.layout.ConversionPattern=%5p %d{HH:mm:ss,SSS} %m%n

# rolling log file
log4j.appender.R=org.apache.log4j.RollingFileAppender
log4j.appender.R.maxFileSize=20MB
log4j.appender.R.maxBackupIndex=50
log4j.appender.R.layout=org.apache.log4j.PatternLayout
log4j.appender.R.layout.ConversionPattern=%5p [%t] %d{ISO8601} %F (line %L) %m%n
# Edit the next line to point to your logs directory
log4j.appender.R.File=/var/log/cassandra/system.log

# Application logging options
#log4j.logger.org.apache.cassandra=DEBUG
#log4j.logger.org.apache.cassandra.db=DEBUG
#log4j.logger.org.apache.cassandra.service.StorageProxy=DEBUG

# Adding this to avoid thrift logging disconnect errors.
log4j.logger.org.apache.thrift.server.TNonblockingServer=ERROR
```

Let's discuss these properties in sequence

- Properties with prefix `log4j.appender.stdout` are for console logging.

- Server logs are generated and appended on a location defined as property `log4j.appender.R.File` value. The default value is `/var/log/cassandra/system`. User can overwrite the property file for default location

- `og4j.appender.R.maxFileSize` defines the maximum log file size.

- The `log4j.appender.R.maxBackupIndex` property defines the maximum rolling log file (default 50).

- The `Log4j.appender.R.layout.ConversionPattern` property defines logging pattern for log files.

- Last line in the `log4j-server.properties` file is for application logging in case of thrift connection with Cassandra. By default it's commented out to avoid unnecessary logging on frequent socket disconnection.

Application Logging Options

By default, Cassandra API level logging is disabled. But we can enable and change log level to log more application level information. Many times applications may need to enable Cassandra-specific server-side logging to troubleshoot the problems. The following code depicts the section that can be used for application-specific logging.

```
# Application logging options
#log4j.logger.org.apache.cassandra=DEBUG
#log4j.logger.org.apache.cassandra.db=DEBUG
#log4j.logger.org.apache.cassandra.service.StorageProxy=DEBUG
```

Changing Log Properties

There are two possible ways for configuring log properties. First, we can modify `log4j-server.properties` and second, via JMX (Java Management Extension), using jconsole. The difference between both of them is, using the latter can change the logging level dynamically at run time, while the first one is static.

Managing Logs via JConsole

JConsole is a GUI monitoring tool for resource usage and performance monitoring of running Java applications using JMX.

The `jconsole` executable can be found in `JDK_HOME/bin`, where `JDK_HOME` is the directory in which the Java Development Kit (JDK) is installed. If this directory is in your system path, you can start JConsole by simply typing `jconsole` at command (shell) prompt. Otherwise, you have to use the full path of the executable file.

On running jconsole, you need to connect the Cassandra Daemon thread as shown in Figure 1-6.

Figure 1-6. *JConsole connection layout*

After successfully connecting to CassandraDaemon process, click on the MBeans tab to look into registered message beans. Figure 1-7 depicts changing the log level for classes within the org.apache.cassandra.db package to INFO level.

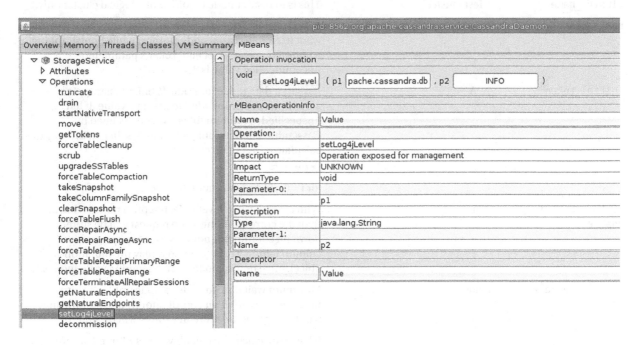

Figure 1-7. *Changing the log level via jconsole Mbeans setting*

■ **Note** Please refer to http://logging.apache.org/log4j/1.2/apidocs/org/apache/log4j/PatternLayout.html for more information on logging patterns.

Understanding Cassandra Configuration

The primary Cassandra configuration file is Cassandra.yaml, which is available within the $CASSSANDRA_HOME/conf folder. Roughly there are approximately 100 properties. Table 1-1 consists of a subset of such properties, which are helpful for Cassandra beginners and worth mentioning.

Table 1-1. *Important Cassandra server properties*

Property	Default	Description
cluster_name	"Test cluster"	This is to restrict node to join in one logical cluster only.
num_tokens	Disabled, not specified	If not specified, default value is 1. For example, if you want to enable virtual node support while bootstrapping a node, you need to set num_tokens parameter. Recommended value is 256.
initial_token	N/A	Assigns a data range for node. While bootstrapping a node, it is recommended to assign a value. If left unspecified, a random token will be assigned by Cassandra. For Random partitioning schema, the way to calculate a initial_token is: $i * (2^{**}127 / N)$ for $i = 0 .. N-1$. N is number of nodes.
hinted_handoff_ enabled	True	With a consistency level ANY, if replica node is down, then the corresponding write request will be stored down on coordinator node as a hint in system.hints column family. This is used to replay the mutation object, once replica node starts accepting write requests.
max_hint_window_in_ ms	3 hours	Maximum wait time for a dead node until new hints meant to be written on coordinator node. After the hint window expires, no more new hints will be stored.
		If left unspecified, the default value is 1 hour. This property is used when writing hints on the coordinator node. If the gossip protocol end point down time for a specific replica node is greater than the specified Maximum wait time value, then no new hints can be written by the StorageProxy service on the coordinator node.
hinted_handoff_ throttle_in_kb	1024	kb/sec hint data flow/per thread.
max_hints_delivery_ threads	2	Maximum number of allowed threads to send data hints. Useful when writing hints across multiple data centers.
populate_io_cache_ on_flush	False	Set it to true, if complete data on a node can fit into memory. Since Cassandra 1.2.2, we can also set this parameter per column family as well. https://issues. apache.org/jira/browse/CASSANDRA-4694.
authenticator	AllowAllAuthenticator	Implementation of IAuthenticator interface. By default Cassandra offers AllowAllAuthenticator and PasswordAuthenticator as internal authentication implementations. PasswordAuthenticator validates username and password against data stored in credentials and users column family in system_auth keyspace. (Security in Cassandra will be discussed at length in Chapter 10.)

(continued)

Table 1-1. (*continued*)

Property	Default	Description
Authorizer	AllowAllAuthorizer	Implementation of IAuthorizer interface. Implementation manages user's permission over keyspace, column family, index, etc. Enabling CassandraAuthorizer on server startup will create a permissions table in system_auth keyspace and to store user permissions.
		(Security in Cassandra will be discussed at length in Chapter 10.)
permissions_validity_in_ms	Default is 2000. Disabled if authorizer property is AllowAllAuthorizer	Default permissions cache validity.
Partitioner	Murmur3Partitioner	Rows distribution across nodes in cluster is decided based on selection partitioner. Available values are RandomPartitioner, ByteOrderedPartitioner, Murmur3Partitioner and OrderPreservingPartitioner (deprecated).
data_file_directories	/var/lib/cassandra/data	Physical data location of node.
commitlog_directory	/var/lib/cassandra/commitlog	Physical location of commit log files of node.
disk_failure_policy	Stop	Available values are stop, best_effort, and ignore. Stop will shut down all communication with node (except JMX). best_effort will still acknowledge read request from available sstables.
key_cache_size_in_mb	Empty, means 100MB or 5% of available heap size, whichever is smaller	To disable set it to Zero.
saved_caches_directory	/var/lib/cassandra/saved_caches	Physical location for saved cache on node.
key_cache_save_period	14400	Key cache save duration (in seconds) save under saved_caches_directory.
key_cache_keys_to_save	Disabled.	By default disabled. All row keys will be cached.
row_cache_size_in_mb	0(Disabled)	In-memory row cache size.
row_cache_save_period	0(Disabled)	row cache save duration (in seconds) save under saved_caches_directory.
row_cache_keys_to_save	Disabled.	By default disabled. All row keys will be cached.

(*continued*)

Table 1-1. (*continued*)

Property	Default	Description
row_cache_provider	SerializingCacheProvider	Available values are SerializingCacheProvider and ConcurrentLinkedHashCacheProvider. SerializingCacheProvider is recommended in case workload is not intensive update as it uses native memory (not JVM) for caching.
commitlog_sync	Periodic	Available values are periodic and batch. In case of batch sync, writes will not be acknowledged until writes are synced with disk. See the commitlog_sync_batch_window_in_ms property.
commitlog_sync_batch_window_in_ms	50	If commitlog_sync is in batch mode, Cassandra will acknowledge writes only after commit log sync windows expires and data will be fsynced to disk.
commitlog_sync_period_in_ms	10000	If commitlog_sync is periodic. Commit log will be fsynced to disk after this value.
commitlog_segment_size_in_mb	32	Commit log segment size. Upon reaching this limit, Cassandra flushes memtables to disk in form of sstables. Keep it to minimum in case of 32 bit JVM to avoid running out of address space and reduced commit log flushing.
seed_provider	SimpleSeedProvider	Implementation of SeedProvider interface. SimpleSeedProvider is default implementation and takes comma separated list of addresses. Default value for "-seeds" parameter is 127.0.0.1. Please change it for multiple node addresses, in case of multi-node deployment.
concurrent_reads	32	If workload data cannot fit in memory, it would require to fetch data from disk. Set this parameter to perform number of concurrent reads.
concurrent_writes	32	Generally writes are faster than reads. So we can set this parameter on the higher side in comparison to concurrent_reads.
memtable_total_space_in_mb	One third of JVM heap(disabled)	Total space allocated for memtables. Once exceeding specified size Cassandra will flush the largest memtable first onto disk.
commitlog_total_space_in_mb	32(32 bit JVM), 1024 (64bit JVM)	Total space allocated commit log segments. Upon reaching the specified limit, Cassandra flushes memtables to claim space by removing the oldest commit log first.
storage_port	7000	TCP port for internal communication between nodes.
ssl_storage_port	7001	Used if client_encryption_options is enabled.
listen_address	Localhost	Address to bind and connect with other Cassandra nodes.

(*continued*)

Table 1-1. (*continued*)

Property	Default	Description
broadcast_address	Disabled(same as listen_address)	Broadcast address for other Cassandra nodes.
internode_authenticator	AllowAllInternode Authenticator	IinternodeAuthenticator interface implementation for internode communication.
start_native_transport	False	CQL native transport for clients.
native_transport_port	9042	CQL native transport port to connect with clients.
rpc_address	Localhost	Thrift rpc address, client to connect with.
rpc_port	9160	Thrift rpc port for clients to communicate.
rpc_min_threads	16	Minimum number of thread for thrift rpc.
rpc_max_threads	2147483647(Maximum 32-bit signed integer)	Maximum number of threads for thrift rpc.
rpc_recv_buff_size_in_bytes	Disabled	Enable if you want to set a limit of receiving socket buffer size for thrift rpc.
rpc_send_buff_size_in_bytes	Disabled	Enable if you want to set a limit of sending socket buffer size for thrift rpc.
incremental_backups	False	If enabled, Cassandra will hard links flushed sstables to backup directory under data_file_directories/keyspace/backup directory.
snapshot_before_compaction	False	If enabled, will create snapshots before each compaction under the data_file_directories/keyspace/snapshots directory.
auto_snapshot	True	If disabled, snapshot will not be taken in case of dml operation (truncate, drop) over keyspace.
concurrent_compactors	Equals number of processors	Equal to cassandra.available_processors (if defined) else number of available processors.
multithreaded_compaction	False	If enabled, single thread per processor will be used for compaction.
compaction_throughput_mb_per_sec	16	Data compaction flow in megabytes per seconds. More compaction throughput will ensure less sstables and more space on disk.
endpoint_snitch	SimpleSnitch	A very important configuration. Snitch can also be termed as informer. Useful to route requests for replica nodes in cluster. Available values are SimpleSnitch, PropertyFileSnitch, RackInferringSnitch, Ec2Snitch, and Ec2MultiRegionSnitch. (I will cover snitch configuration in later chapters.)

(*continued*)

Table 1-1. (*continued*)

Property	Default	Description
request_scheduler	NoScheduler	Client request scheduler. By default no scheduling is done, but we can configure this to RoundRobinScheduler or a custom implementation. It will queue up client dml request and finally release it after successfully processing the request.
server_encryption_ options	None	To enable encryption for internode communication. Available values are all, none, dc, and rack.
client_encryption_ options	false(not enabled)	To enable client/server communication. If enabled must specify ssl_storage_port. As it will be used for client/server communication.
internode_ compression	All	To compress traffic in internode communication. Available values are: all, dc, and none.
inter_dc_tcp_nodelay	True	Setting it to false will cause less congestion over TCP protocol but increased latency.

Commit Log Archival

To enable Cassandra for auto commit log archiving and restore for recovery (supported since 1.1.1.), the commitlog_archiving.properties file is used. It configures archive_command and restore_command properties. Commit log archival is also referred to as write ahead log (WAL) archive and used for point-in-time recovery.

Cassandra's implementation is similar to Postgresql. Postgresql is an object-oriented relational database management system (OORDBMS) that offers wal_level settings with minimum as the lowest, followed by archive and hot_standby levels to allow executing queries during recovery. For more details on Postgresql refer to http://www.postgresql.org/.

archive_command

Enable archive_command for implicit commit log archival using a command such as:

```
archive_command= /bin/ln %path /home/backup/%name
```

Here %path is a fully qualified path of the last active commit log segment and %name is the name of commit log. The above-mentioned shell command will create a hard link for the commit log segment (%path). If row mutation size exceeds commitlog_segment_size_in_mb, Cassandra archives this segment using the archive command under /home/backup/. Here %path is the name of latest old segment and %name is commit log file name.

restore_command

Leaving restore_command and restore_directories blank in commitlog_archiving.properties during bootstrap Cassandra will replay these log files using the restore_command:

```
restore_command=cp -f %from %to
```

Here %from is a value specified as restore_directories and %to is next commit log segment file under commitlog_directory.

One advantage of this continuous commit log is high availability of data also termed warm standby.

Configuring Replication and Data Center

Recently, the need for big data heterogeneous systems has evolved. Components in such systems are diverse in nature and can be made up of different data sets. Considering nature, locality, and quantity of data volume, it is highly possible that such systems may need to interconnect with data centers available on different physical locations.

A data center is a hardware system (say commodity server), which consists of multiple racks. A rack may contain one or more nodes (see Figure 1-8).

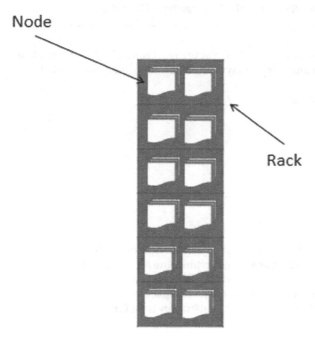

Figure 1-8. *Image depicting a Cassandra data center*

Reasons for maintaining multiple data centers can be high availability, stand-by-node, and data recovery.

With high availability, any incoming request must be served with minimum latency. Data replication is a mechanism to keep redundant copy of the same data on multiple nodes.

As explained above, a data center consists of multiple racks with each rack containing multiple nodes. A data replication strategy is vital in order to enable high availability and node failure. Situations like

- Local reads (high availability)

- Fail-over (node failure)

Considering these factors, we should replicate data on multiple nodes in the same data center but with different racks. This would avoid read/write failure (in case of network connection issues, power failure, etc.) of nodes in the same rack.

Replication means keeping redundant copies of data over multiple data nodes for high availability and consistency. With Cassandra we can configure the replication factor and replication strategy class while creating keyspace.

While creating schema (rather than keyspace) we can configure replication as:

```
CREATE KEYSPACE apress WITH replication = {'class': 'SimpleStrategy', 'replication_factor' : 3};
// cql3 script

create keyspace apress with placement_strategy='org.apache.cassandra.locator.SimpleStrategy' and
strategy_options ={replication_factor:1};
// using cassandra-cli thrift
```

■ **Note**   Schema creation and management via CQL3 and Cassandra-cli will be discussed in Chapter 2.

Here, SimpleStrategy is the replication strategy, where the replication_factor is 3. Using SimpleStrategy like this, each data row will be replicated on 3 replica nodes synchronously or asynchronously (depending on the write consistency level) in clockwise direction.

Different strategy class options supported by Cassandra are

- SimpleStrategy
- LocalStrategy
- NetworkTopologyStrategy

LocalStrategy

LocalStrategy is available for internal purposes and is used for system and system_auth keyspaces. System and system_auth are internal keyspaces, implicitly handled by Cassandra's storage architecture for managing authorization and authentication. These keyspaces also keep metadata about user-defined keyspaces and column families. In the next chapter we will discuss them in detail. Trying to create keyspace with strategy class as **LocalStrategy** is not permissible in Cassandra and would give an error like "LocalStrategy is for Cassandra's internal purpose only".

NetworkTopologyStrategy

NetworkTopologyStrategy is preferred if multiple replica nodes need to be placed on different data centers. We can create a keyspace with this strategy as

```
CREATE KEYSPACE apress WITH replication = {'class': 'NetworkTopologyStrategy', 'dc1' : 2, 'dc2' : 3};
```

Here dc1 and dc2 are data center names with replication factor of 2 and 3 respectively. Data center names are derived from a configured snitch property.

SimpleStrategy

SimpleStrategy is recommended for multiple nodes over multiple racks in a single data center.

```
CREATE KEYSPACE apress WITH replication = {'class': 'SimpleStrategy', 'replication_factor' : 3};
```

Here, replication factor 3 would mean to replicate data on 3 nodes and strategy class SimpleStrategy would mean to have those Cassandra nodes within the same data center.

Cassandra Multiple Node Configuration

In this section, we will discuss multiple Cassandra node configurations over a single machine and over Amazon EC2 instances. Reasons to choose AWS EC2 instances include the setup of the Cassandra cluster over the cloud and the set up on the local box to configure the Cassandra cluster over physical boxes. AWS based configuration would educate users about AWS and Cassandra.

Configuring Multiple Nodes over a Single Machine

Configuring multiple nodes over a single machine is more of an experiment, as with a production application you would like to configure a Cassandra cluster over multiple Cassandra nodes. Setting up multinode clusters over a single machine or multiple machines is similar. That's what we will be covering in this sample exercise. In this example, we will configure 3 nodes (127.0.0.2-4) on a single machine.

1. We need to map hostnames to IP addresses.

    a. In Windows and Linux OS, these configurations are available in etc/hosts (Windows) or /etc/hosts (Linux) files. Modify the configuration file to add the above-mentioned 3 node configuration as:

    ```
    127.0.0.1   127.0.0.2
    127.0.0.1   127.0.0.3
    127.0.0.1   127.0.0.4
    ```

    b. For Mac OS, we need to create those aliases as:

    ```
    sudo ifconfig lo0 alias 127.0.0.2 up
    sudo ifconfig lo0 alias 127.0.0.3 up
    sudo ifconfig lo0 alias 127.0.0.4 up
    ```

2. Unzip the downloaded Cassandra tarball installation in 3 different folders (one for each node). Assign each node an identical cluster_name as:

    ```
    # The name of the cluster. This is mainly used to prevent machines in
    # one logical cluster from joining another.
    cluster_name: 'Test Cluster'
    ```

3. We should hold identical seeds on each node in the cluster. These are used just to initiate gossip protocol among nodes in the cluster. Configure seeds in cassandra.yaml as :

```
seed_provider:
    # Addresses of hosts that are deemed contact points.
    # Cassandra nodes use this list of hosts to find each other and learn
    # the topology of the ring.  You must change this if you are running
    # multiple nodes!
    - class_name: org.apache.cassandra.locator.SimpleSeedProvider
      parameters:
          # seeds is actually a comma-delimited list of addresses.
          # Ex: "<ip1>,<ip2>,<ip3>"
          - seeds: "127.0.0.2"
```

4. Change the listen_address and rpc_address configurations for 127.0.0.2, 127.0.0.3, and 127.0.0.4 IP addresses in each cassandra.yaml file. Since all 3 nodes are running on the same machine, change the rpc_address to 9160, 9161, and 9162 for each respectively.

5. Here we have an option to choose between 1 token per node or multiple tokens per node. Cassandra 1.2 introduced the "Virtual nodes" feature which allows assigning a range of tokens on a node. We will discuss Virtual nodes in coming chapter. Change the initial_token to empty and keep num_tokens as 2 (recommend is 256).

6. Next is to assign different JMX_PORT (say 8081, 8082, and 8083) for each node.

   a. With Linux, modify $CASSANDRA_HOME/conf/cassandra.env.sh as:

   ```
   # Specifies the default port over which Cassandra will be available for
   # JMX connections.
   JMX_PORT="7199"
   ```

   b. With Windows, modify $CASSANDRA_HOME/bin/cassandra.bat as:

   ```
   REM ***** JAVA options *****
   set JAVA_OPTS=-ea^
    -javaagent:"%CASSANDRA_HOME%\lib\jamm-0.2.5.jar"^
    -Xms1G^
    -Xmx1G^
    -XX:+HeapDumpOnOutOfMemoryError^
    -XX:+UseParNewGC^
    -XX:+UseConcMarkSweepGC^
    -XX:+CMSParallelRemarkEnabled^
    -XX:SurvivorRatio=8^
    -XX:MaxTenuringThreshold=1^
    -XX:CMSInitiatingOccupancyFraction=75^
    -XX:+UseCMSInitiatingOccupancyOnly^
    -Dcom.sun.management.jmxremote.port=7199^
    -Dcom.sun.management.jmxremote.ssl=false^
    -Dcom.sun.management.jmxremote.authenticate=false^
    -Dlog4j.configuration=log4j-server.properties^
    -Dlog4j.defaultInitOverride=true
   ```

7. Let's start each node one by one and check ring status as: $CASSANDRA_HOME/apache-cassandra-1.2.4/bin/nodetool -h 127.0.0.02 -p 8081 ring.

Figure 1-9 shows ring status while connecting to one Cassandra node using jmx. Since Cassandra's architecture is peer-to-peer, checking ring status on any node will yield the same result.

```
Datacenter: datacenter1
==========
Replicas: 1

Address    Rack    Status State    Load       Owns      Token
                                                         6462923463760331336
127.0.0.4  rack1   Up     Normal   71.77 KB   41.80%    3591055736266366120
127.0.0.4  rack1   Up     Normal   71.77 KB   41.80%    943929740184731686
127.0.0.2  rack1   Up     Normal   74.62 KB   45.50%    -6767292597249134115
127.0.0.2  rack1   Up     Normal   74.62 KB   45.50%    4485908808910730573
127.0.0.3  rack1   Up     Normal   51.45 KB   12.70%    13089258224509809020
127.0.0.3  rack1   Up     Normal   51.45 KB   12.70%    6462923463760331336
```

Figure 1-9. *The ring status*

Configuring Multiple Nodes over Amazon EC2

Amazon Elastic Computing Cloud (Amazon EC2), one of the important parts of Amazon Web Service (AWS) cloud computing platform. AWS offers you to choose OS platform and provides required hardware support over the cloud, which allows you to quickly set up and deploy application over the cloud computing platform. To learn more about Amazon ec2 setup please refer to http://aws.amazon.com/ec2/.

In this section, we will learn about how to configure multiple Cassandra nodes over Amazon EC2. To do so, follow these steps.

1. First let's launch 2 instances of AMI (ami-00730969), as shown in Figure 1-10.

	Name	Instance	AMI ID	Root Device	Type	State	Status Checks	Alarm Status	Monitoring	Security Groups	Key
	ec2instance2	i-58dc6c39	ami-00730969	ebs	t1.micro	running	2/2 checks p:	none	basic	viveksecgrp	vivek
	ec2instance1	i-5edc6c3f	ami-00730969	ebs	t1.micro	running	2/2 checks p:	none	basic	viveksecgrp	vivek

Viewing: All Instances | All Instance Types | Search | 1 to 2 of 2 Instances

Figure 1-10. *Ec2 console display with 2 instances in running state*

2. Modify security group to enable 9160, 7000, and 7199 ports, as in Figure 1-11.

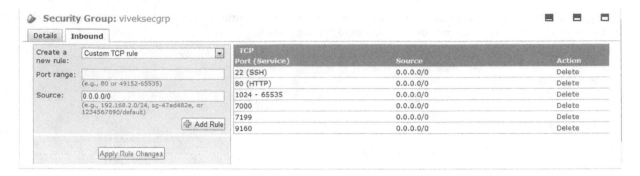

***Figure 1-11.** Configuring security group settings*

3. Connect to each instance and download Cassandra tarball as:

   ```
   wget http://archive.apache.org/dist/cassandra/1.2.4/apache-cassandra-1.2.4-bin.tar.gz
   ```

4. Download and setup Java on each EC2 instance using the rpm installer as:

   ```
   sudo rpm -i jdk-7-linux-x64.rpm
   sudo rm -rf /usr/bin/java
   sudo ln -s /usr/java/jdk1.7.0/bin/java /usr/bin/java
   sudo rm -rf /usr/bin/javac
   sudo ln -s /usr/java/jdk1.7.0/bin/javac /usr/bin/javac
   ```

5. Multiple Cassandra node configurations are the same as we discussed in the previous section. In this section we will demonstrate using single token per node (initial_token). Let's assign initial token values 0 and 1. We can assign initial_token values by modifying Cassandra.yaml files on each node.

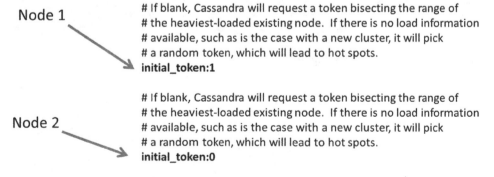

***Figure 1-12.** initial_token configuration for both nodes*

6. Create any one of these two as a seed node and keep storage port, jmx_port, and rpc_port to 7000, 7199, and 9160.

7. Let's keep listen_address and rpc_address with empty values (default is the node's inet address (underlined), shown in Figure 1-13).

```
[ec2-user@ip-10-145-213-3 ~]$ ifconfig
eth0      Link encap:Ethernet  HWaddr 22:00:0A:91:D5:03
          inet addr:10.145.213.3  Bcast:10.145.213.63  Mask:255.255.255.192
          inet6 addr: fe80::2000:aff:fe91:d503/64 Scope:Link
          UP BROADCAST RUNNING MULTICAST  MTU:1500  Metric:1
          RX packets:2014 errors:0 dropped:0 overruns:0 frame:0
          TX packets:1908 errors:0 dropped:0 overruns:0 carrier:0
          collisions:0 txqueuelen:1000
          RX bytes:204991 (200.1 KiB)  TX bytes:195399 (190.8 KiB)
          Interrupt:25

lo        Link encap:Local Loopback
          inet addr:127.0.0.1  Mask:255.0.0.0
          inet6 addr: ::1/128 Scope:Host
          UP LOOPBACK RUNNING  MTU:16436  Metric:1
          RX packets:0 errors:0 dropped:0 overruns:0 frame:0
          TX packets:0 errors:0 dropped:0 overruns:0 carrier:0
          collisions:0 txqueuelen:0
          RX bytes:0 (0.0 b)  TX bytes:0 (0.0 b)
```

Figure 1-13. How to get inet address for node

8. Let's start each node one by one and check ring status. Verify both EC2 instances should be up, running, and connected using ring topology. Figure 1-14 shows the ring status of both running ec2 instances.

```
[ec2-user@ip-10-145-213-3 ~]$ software/apache-cassandra-1.2.4/bin/nodetool ring

Datacenter: datacenter1
==========
Replicas: 1

Address         Rack      Status State   Load       Owns        Token
                                                                 1
10.145.213.3    rack1     Up     Normal  55.24 KB   100.00%     0
10.167.12.16    rack1     Up     Normal  36.57 KB   0.00%       1
```

Figure 1-14. The two EC2 instances and their ring statuses

9. Figure 1-15 shows instance 10.145.213.3 is up and joining the cluster ring.

```
INFO 17:25:49,677 Completed flushing /var/lib/cassandra/data/system/local/system-local-ib-1-Data.db (350 bytes) for commitlog po
148759, position=50656)
INFO 17:25:49,810 Starting Messaging Service on port 7000
INFO 17:25:49,889 Enqueuing flush of Memtable-local@599262429(86/86 serialized/live bytes, 4 ops)
INFO 17:25:49,890 Writing Memtable-local@599262429(86/86 serialized/live bytes, 4 ops)
INFO 17:25:49,921 Completed flushing /var/lib/cassandra/data/system/local/system-local-ib-2-Data.db (122 bytes) for commitlog po
148759, position=50933)
INFO 17:25:49,923 JOINING: waiting for ring information
INFO 17:25:50,879 Node /10.145.213.3 is now part of the cluster
INFO 17:25:50,880 InetAddress /10.145.213.3 is now UP
INFO 17:25:50,904 Enqueuing flush of Memtable-peers@288389262(73/73 serialized/live bytes, 4 ops)
INFO 17:25:50,905 Writing Memtable-peers@288389262(73/73 serialized/live bytes, 4 ops)
INFO 17:25:50,946 Completed flushing /var/lib/cassandra/data/system/peers/system-peers-ib-1-Data.db (136 bytes) for commitlog po
148759, position=51229)
INFO 17:25:50,990 Enqueuing flush of Memtable-schema_keyspaces@1386215844(389/389 serialized/live bytes, 11 ops)
INFO 17:25:50,991 Writing Memtable-schema_keyspaces@1386215844(389/389 serialized/live bytes, 11 ops)
INFO 17:25:51,025 Completed flushing /var/lib/cassandra/data/system/schema_keyspaces/system-schema_keyspaces-ib-1-Data.db (262 b
tion(segmentId=1375205148759, position=53711)
```

Figure 1-15. *Node 10.145.213.3 is up and joining the cluster*

Summary

This chapter is an introductory one to cover all generic concepts and Cassandra-specific configurations. For application developers it is really important to understand the essence of replication, data distribution, and most importantly setting this up with Cassandra. Now we are ready for the next challenge: handling big data with Cassandra! In the next chapter we will discuss Cassandra's storage mechanism and data modeling. With data modeling and understanding Cassandra's storage architecture it would help us to model the data set, and analyze and look into the best possible approaches available with Cassandra.

CHAPTER 2

■ ■ ■

Cassandra Data Modeling

In the previous chapter we discussed Cassandra configuration, installation, and cluster setup. This chapter will walk you through

- Data modeling concepts
- Cassandra collection support
- CQL vs thrift based schema
- Managing data types
- Counter columns

Get ready to learn with an equal balance of theoretical and practical approach.

Introducing Data Modeling

Data modeling is a mechanism to define read/write requirements and build a logical structure and object model. Cassandra is an NOSQL database and promotes **read-before-write** instead of relational model. Read-before-write or ready-for-read design is used to analyze your data read requirement first and store it in the same way. Consider managing data volume in peta bytes or zeta bytes, where we cannot afford to have in-memory computations (e.g., joins) because of data volume. Hence it is preferable to have the data set ready for retrieval or large data analytics. Users need not know about columns up front, but should avoid storing flat columns and favor doing computations (e.g., aggregation, joins etc.) during read time.

Cassandra is a column-family–oriented database. Column family, as the name suggests is "family of columns." Each row in Cassandra may contain one or more columns. A column is the smallest unit of data containing a name, value, and time stamp (see Figure 2-1).

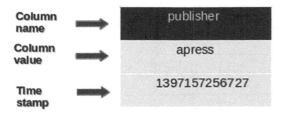

Figure 2-1. *Cassandra column definition*

By default Cassandra distribution comes with cqlsh and Cassandra-cli command line clients to manipulate. **Cassandra-cli** and **cqlsh** (.sh and .bat) are available under bin folder. Running these command line clients over Linux, Windows, or Mac is fairly easy. Running shell files over Linux and Mac box requires simply running cql.sh. However running cqlsh over Windows would require Python to be installed.

To install cqlsh on Windows, follow these steps:

1. First, download Python from https://www.python.org/ftp/python/2.7.6/python-2.7.6.msi.

2. Add python.exe to PATH under environment variable

3. Run setup.py, available under $CASSANDRA_HOME/pylib directory:

   ```
   python setup.py install
   ```

4. Run cqlsh, available under bin directory (see Figure 2-2):

   ```
   python cqlsh
   ```

```
C:\vivek\apache-cassandra-1.2.4-bin\apache-cassandra-1.2.4\bin>python cqlsh
Connected to Test Cluster at localhost:9160.
[cqlsh 2.3.0 | Cassandra 1.2.4 | CQL spec 3.0.0 | Thrift protocol 19.35.0]
Use HELP for help.
cqlsh>
cqlsh>
```

Figure 2-2. *successfully connected to cql shell*

Data Types

Before CQL's evolution, data types in Cassandra are defined in the form of a comparator and validator. Column or row key value is referred to as a *validator*, whereas a column name is called a *comparator*. Available data types are shown in Figure 2-3.

Internal Type	CQL Name	Description
BytesType	blob	Arbitrary hexadecimal bytes (no validation)
AsciiType	ascii	US-ASCII character string
UTF8Type	text, varchar	UTF-8 encoded string
IntegerType	varint	Arbitrary-precision integer
LongType	int, bigint	8-byte long
UUIDType	uuid	Type 1 or type 4 UUID
DateType	timestamp	Date plus time, encoded as 8 bytes since epoch
BooleanType	boolean	true or false
FloatType	float	4-byte floating point
DoubleType	double	8-byte floating point
DecimalType	decimal	Variable-precision decimal
CounterColumnType	counter	Distributed counter value (8-byte long)

Figure 2-3. *Cassandra's supported data types*

Dynamic Columns

Since its inception, Cassandra is projected as a schema-less, column-family–oriented distributed database. The number of columns may vary for each row in a column family. A column definition can be added dynamically at run time.

Cassandra-cli (Thrift) and cqlsh (CQL3) are two command clients we will be using for various exercises in this chapter.

Dynamic Columns via Thrift

Let's discuss a simple Twitter use case. In this example we would explore ways to model and store dynamic columns via Thrift.

1. First, let's create a keyspace **twitter** and column family **users**:

    ```
    create keyspace twitter with strategy_options={replication_factor:1} and
    placement_strategy='org.apache.cassandra.locator.SimpleStrategy';
    use twitter;
    create column family users with key_validation_class='UTF8Type' and
    comparator='UTF8Type' and default_validation_class='UTF8Type';
    ```

    Here, while defining a column family, we did not define any columns with the column family. Columns will be added on the fly against each row key value.

2. Store a few columns in the **users** column family for row key value **'imvivek'**:

    ```
    set users['imvivek']['apress']='apress author';
    set users['imvivek']['team_marketing']='apress marketing';
    set users['imvivek']['guest']='guest user';
    set users['imvivek']['ritaf']='rita fernando';
    ```

    Here we are adding followers as dynamic columns for user imvivek.

3. Let's add **'imvivek'** and **'team_marketing'** as followers for **'ritaf'**:

    ```
    set users['ritaf']['imvivek']='vivek mishra';
    set users['team_marketing']['imvivek']='vivek mishra';
    ```

4. To view a list of rows in **users** column family (see Figure 2-4), use the following command:

    ```
    list users;
    ```

```
[default@twitter] list users;
Using default limit of 100
Using default cell limit of 100
-------------------
RowKey: team_marketing
=> (name=imvivek, value=vivek mishra, timestamp=1397492166986000)
-------------------
RowKey: ritaf
=> (name=imvivek, value=vivek mishra, timestamp=1397492153545000)
-------------------
RowKey: imvivek
=> (name=apress, value=apress author, timestamp=1397491295119000)
=> (name=guest, value=guest user, timestamp=1397492111388000)
=> (name=ritaf, value=rita fernando, timestamp=1397492132782000)
=> (name=team_marketing, value=apress marketing, timestamp=1397492060977000)

3 Rows Returned.
Elapsed time: 13 msec(s).
```

Figure 2-4. *Output of selecting users*

In Figure 2-4, we can see column name and their values against each row key stored in step 3.

5.   We can delete columns for an individual key as well. For example, to delete a column **'apress'** for row key **'imvivek'**:

```
del users['imvivek']['apress'];
```

Figure 2-5 shows the number of columns for imvivek after step 5.

```
[default@twitter] list users;
Using default limit of 100
Using default cell limit of 100
-------------------
RowKey: team_marketing
=> (name=imvivek, value=vivek mishra, timestamp=1397492166986000)
-------------------
RowKey: ritaf
=> (name=imvivek, value=vivek mishra, timestamp=1397492153545000)
-------------------
RowKey: imvivek
=> (name=guest, value=guest user, timestamp=1397492111388000)
=> (name=ritaf, value=rita fernando, timestamp=1397492132782000)
=> (name=team_marketing, value=apress marketing, timestamp=1397492060977000)

3 Rows Returned.
Elapsed time: 11 msec(s).
```

Figure 2-5. *The number of columns for imvivek after deletion*

Here column name is the follower's twitter_id and their full name is column value. That's how we can manage schema and play with dynamic columns in Thrift way. We will discuss dynamic column support with CQL3 in Chapter 3.

Dynamic Columns via cqlsh Using Map Support

In this section, we will discuss how to implement the same Twitter use case using map support. Collection support in Cassandra would work only with CQL3 binary protocol.

1. First, let's create a keyspace **twitter** and column family **users**:

    ```
    create keyspace twitter with replication = {'class':'SimpleStrategy',
    'replication_factor':3};
    use twitter;
    create table users(twitter_id text primary key,followers map<text,text>);
    ```

2. Store a few columns in **users** column family for row key value **'imvivek'**:

    ```
    insert into users(twitter_id,followers) values('imvivek',{'guestuser':'guest',
    'ritaf':'rita fernando','team_marketing':'apress marketing'});
    ```

    Here we are adding followers as dynamic columns as map attributes for user imvivek.

3. Let's add **'imvivek'** and **'team_marketing'** as followers for **'ritaf'**:

    ```
    insert into users(twitter_id,followers) values('ritaf',{'imvivek':'vivek mishra'});
            insert into users(twitter_id,followers) values('team_marketing',
    {'imvivek':'vivek mishra'});
    ```

4. To view list of rows in the **users** column family (see Figure 2-6), use the following command:

    ```
    select * from users;
    ```

```
cqlsh:twitter> select * from users;

 twitter_id     | followers
----------------+-------------------------------------------------------------------------------
 team_marketing |                                                      {'imvivek': 'vivek mishra'}
          ritaf |                                                      {'imvivek': 'vivek mishra'}
        imvivek | {'guestuser': 'guest', 'ritaf': 'rita fernando', 'team_marketing': 'apress marketing'}

(3 rows)
```

Figure 2-6. *Map containing followers for user*

5. To add 'team_marketing' as a follower for 'ritaf' and vice versa (see Figure 2-7), we can simply add it as an element in **users** column family:

    ```
    update users set followers = followers + {'team_marketing':'apress marketing'} where
    twitter_id='ritaf';
    ```

    ```
    update users set followers = followers + {'ritaf':'rita fernando'} where
    twitter_id='apress_marketing';
    ```

```
cqlsh:twitter> update users set followers = followers + {'team_marketing':'apress marketing'} where twitter_id='ritaf';
cqlsh:twitter> update users set followers = followers + {'ritaf':'rita fernando'} where twitter_id='apress_marketing';
cqlsh:twitter> select * from users;

 twitter_id       | followers
------------------+------------------------------------------------------------------------------------------
   team_marketing |                                                              {'imvivek': 'vivek mishra'}
            ritaf |                       {'imvivek': 'vivek mishra', 'team_marketing': 'apress marketing'}
          imvivek | {'guestuser': 'guest', 'ritaf': 'rita fernando', 'team_marketing': 'apress marketing'}
 apress_marketing |                                                             {'ritaf': 'rita fernando'}

(4 rows)
```

Figure 2-7. *After update map of followers for each user*

6. Using update would work as an insert if row key doesn't exist in the database. For example,

```
update users set followers = followers + {'ritaf':'rita fernando'} where
twitter_id='jhassell';  // update as insert
```

Figure 2-8 shows that ritaf has been added as a follower of jhassell.

```
cqlsh:twitter> update users set followers = followers + {'ritaf':'rita fernando'} where twitter_id='jhassell';
cqlsh:twitter> select * from users;

 twitter_id       | followers
------------------+------------------------------------------------------------------------------------------
         jhassell |                                                             {'ritaf': 'rita fernando'}
   team_marketing |                                                              {'imvivek': 'vivek mishra'}
            ritaf |                       {'imvivek': 'vivek mishra', 'team_marketing': 'apress marketing'}
          imvivek | {'guestuser': 'guest', 'ritaf': 'rita fernando', 'team_marketing': 'apress marketing'}
 apress_marketing |                                                             {'ritaf': 'rita fernando'}

(5 rows)
```

Figure 2-8. *Update works as an insert for map of followers for nonexisting row key (e.g., twitter_id)*

7. To delete an element from the map we need to execute, use this command:

```
delete followers['guestuser'] from users where twitter_id='imvivek';
```

You can see that the list of followers for imvivek is reduced to four followers after deletion (see Figure 2-9).

```
cqlsh:twitter> delete followers['guestuser'] from users where twitter_id='imvivek';
cqlsh:twitter> select * from users;

 twitter_id       | followers
------------------+------------------------------------------------------------------------------------------
         jhassell |                                                             {'ritaf': 'rita fernando'}
   team_marketing |                                                              {'imvivek': 'vivek mishra'}
            ritaf |                       {'imvivek': 'vivek mishra', 'team_marketing': 'apress marketing'}
          imvivek |                        {'ritaf': 'rita fernando', 'team_marketing': 'apress marketing'}
 apress_marketing |                                                             {'ritaf': 'rita fernando'}

(5 rows)
```

Figure 2-9. *After deleting guestuser as a follower for imvivek*

With that said, we can add a dynamic column as a key-value pair using collection support.

Dynamic Columns via cqlsh Using Set Support

Consider a scenario where the user wants to store only a collection of follower's id (not full name). Cassandra offers collection support for keeping a **list** or **set** of such elements. Let's discuss how to implement it using **set** support.

1. First, let's create a keyspace **twitter** and column family **users**.

   ```
   create keyspace twitter with replication = {'class':'SimpleStrategy',
   'replication_factor':3};
   use twitter;
   create table users(twitter_id text primary key,followers set<text>);
   ```

2. Store few columns in **users** column family for row key value 'imvivek'.

   ```
   insert into users(twitter_id,followers) values('imvivek',
   {'guestuser','ritaf','team_marketing'});
   ```

   Here we are adding followers as dynamic columns as set attributes for user imvivek.

3. Let's add the following:

   ```
   'imvivek' and 'team_marketing' as followers for 'ritaf'
   'ritaf' as a follower for 'jhassell'

   insert into users(twitter_id,followers) values('ritaf', {'imvivek','jhassell',
   'team_marketing'});
   insert into users(twitter_id,followers) values('jhassell', {'ritaf'});
   ```

4. To view the list of rows in **users** column family (see Figure 2-10), use the following command:

   ```
   select * from users;
   ```

```
cqlsh:twitter> select * from users;

 twitter_id | followers
------------+----------------------------------------
    jhassell |                                {'ritaf'}
       ritaf | {'imvivek', 'jhassell', 'team_marketing'}
     imvivek |   {'guestuser', 'ritaf', 'team_marketing'}

(3 rows)
```

Figure 2-10. *Followers for ritaf, jhassell, and imvivek have been added*

5. We can update the collection to delete an element as follows. Figure 2-10 shows the result:

   ```
   update users set followers = followers - {'guestuser'} where twitter_id = 'imvivek';
   ```

```
cqlsh:twitter> update users set followers = followers - {'guestuser'} where twitter_id = 'imvivek';
cqlsh:twitter> select * from users;

 twitter_id | followers
------------+--------------------------------------------------
   jhassell |                                         {'ritaf'}
      ritaf | {'imvivek', 'jhassell', 'team_marketing'}
    imvivek |                    {'ritaf', 'team_marketing'}

(3 rows)
```

Figure 2-11. *Updated set of followers after removing guestuser for imvivek*

Collection support can be a good alternative for achieving Adding dynamic columns over Cassandra. Composite key is a combination of multiple table fields where the first part is referred to as **partition key** and the remaining part of the composite key is known as cluster key. Chapter 3 will discuss achieving dynamic columns using composite columns.

Secondary Indexes

In a distributed cluster, data for a column family is distributed across multiple nodes, based on replication factor and partitioning schema. However data for a given row key value will always be on the same node. Using the primary index (e.g., Row key) we can always retrieve a row. But what about retrieving it using non-row key values?

Cassandra provides support to add indexes over column values, called **Secondary indexes**. Chapter 3 will cover more about indexes, so for now let's just take a look at a simple secondary index example.

Let's discuss the same Twitter example and see how we can utilize and enable secondary index support.

1. First, let's create **twitter** keyspace and column family **users**.

    ```
    create keyspace twitter with replication = {'class' : 'SimpleStrategy' ,
    'replication_factor' : 3};

    use twitter;

    create table users(user_id text PRIMARY KEY,fullname text,email text,password text,
    followers map<text, text>);
    ```

2. Insert a user with e-mail and password:

    ```
    insert into users(user_id,email,password,fullname,followers) values ('imvivek',
    'imvivek@xxx.com','password','vivekm',{'mkundera':'milan kundera','guest': 'guestuser'});
    ```

Before we move ahead with this exercise, it's worth discussing which columns should be indexed?

Any read request using the secondary index will actually be broadcast to all nodes in a cluster. Cassandra maintains a hidden column family for the secondary index locally on node, which is scanned for retrieving rows using secondary indexes.

While performing data modeling, we should create secondary indexes over column values which should return a big chunk of data over a very large data set. Indexes over unique values of small data sets would simply become an overhead, which is not a good data modeling practice. Index over fullname is a possible candidate for indexing.

3. Let's create secondary index over fullname

    ```
    create index fullname_idx on users(fullname);
    ```

After successful index creation, we can fetch records using fullname. Figure 2-12 shows the result.

```
cqlsh:twitter> select * from users where fullname='vivekm';

 user_id | email                 | followers                                    | fullname | password
---------+-----------------------+----------------------------------------------+----------+----------
 imvivek | imvivek@outlook.com   | {guest: guestuser, mkundera: milan kundera}  |   vivekm | password
```

Figure 2-12. Search user for records having fullname value vivekm

4. Let's add a column of **age** and create the index:

```
alter table users add age text;
create index age_idx on users(age);
update users set age='32' where user_id='imvivek';
insert into users(user_id,email,password,fullname,followers,age) values
('mkundera','mkundera@outlook.com','password','milan kundera',{'imvivek':'vivekm','gues
t': 'guestuser'},'51');
```

Figure 2-13 shows the outcome.

```
cqlsh:twitter> select * from users where age='51';

 user_id  | age | email                 | followers                              | fullname      | password
----------+-----+-----------------------+----------------------------------------+---------------+----------
 mkundera |  51 | mkundera@outlook.com  | {guest: guestuser, imvivek: vivekm}    | milan kundera | password
```

Figure 2-13. Selecting all users of age 51

5. Let's alter data type of age to int:

```
alter table users alter age type int;
```

It will result in the following error:

```
TSocket read 0 bytes (via cqlsh)
```

6. To alter data type of indexed columns we need to rebuild them:

```
drop index age_idx;
alter table users alter age type int;
```

But please note that in such cases, it may result the data set being in an incompatible state (see Figure 2-14).

```
cqlsh:twitter> select * from users;

 user_id | age | email                | followers                              | fullname      | password
---------+-----+----------------------+----------------------------------------+---------------+---------
mkundera | '51'| mkundera@outlook.com |      {guest: guestuser, imvivek: vivekm}| milan kundera | password
 imvivek | '32'| imvivek@outlook.com  | {guest: guestuser, mkundera: milan kundera}|       vivekm | password

Failed to decode value '51' (for column 'age') as int: unpack requires a string argument of length 4
Failed to decode value '32' (for column 'age') as int: unpack requires a string argument of length 4
```

Figure 2-14. *Error while changing data type to int from string*

Here is the error:

```
Failed to decode value '51' (for column 'age') as int: unpack requires a string argument of length 4
Failed to decode value '32' (for column 'age') as int: unpack requires a string argument of length 4
```

Hence it is recommended to change data types on indexed columns, when there is no data available for that column.

Indexes over collections are not supported in Cassandra 2.0. Figure 2-15 shows what happens if we try to create an index follower. However, before this book went to press, version 2.1 was released and added this capability. See "Indexing on Collection Attributes" in Chapter 11.

```
cqlsh:twitter> create index followers_idx on users(followers);
Bad Request: Indexes on collections are no yet supported
```

Figure 2-15. *Indexes over collections are not supported in Cassandra 2.0*

■ **Note**   Updates to the data type of clustering keys and indexes are not allowed.

CQL3 and Thrift Interoperability

Prior to CQL existence, Thrift was the only way to develop an application over Cassandra. CQL3 and Thrift interoperability issues are often discussed within the Cassandra community.

Let's discuss some issues with a simple example:

1. First, let's create a keyspace and column family using CQL3.

    ```
    create keyspace cql3usage with replication = {'class' : 'SimpleStrategy' ,
    'replication_factor' : 3};
    use cql3usage;
    create table user(user_id text PRIMARY KEY, first_name text, last_name text,
    emailid text);
    ```

2. Let's insert one record:

    ```
    insert into user(user_id,first_name,last_name,emailid)
    values('@mevivs','vivek','mishra','vivek.mishra@xxx.com');
    ```

3. Now, connect with Cassandra-cli (the Thrift way) and update the **user** column family to create indexes over last_name and first_name:

```
update column family user with key_validation_class='UTF8Type' and column_
metadata=[{column_name:last_name, validation_class:'UTF8Type', index_type:KEYS},
{column_name:first_name, validation_class:'UTF8Type', index_type:KEYS}];
```

■ **Note**   Chapter 3 will cover indexing in detail.

---

4. Now explore the **user** column family with CQL3, and see the result in Figure 2-16.

```
describe table user;
```

```
CREATE TABLE user (
  key text PRIMARY KEY
) WITH
  bloom_filter_fp_chance=0.010000 AND
  caching='KEYS_ONLY' AND
  comment='' AND
  dclocal_read_repair_chance=0.000000 AND
  gc_grace_seconds=864000 AND
  read_repair_chance=0.100000 AND
  replicate_on_write='true' AND
  populate_io_cache_on_flush='false' AND
  compaction={'class': 'SizeTieredCompactionStrategy'} AND
  compression={'sstable_compression': 'SnappyCompressor'};
```

Figure 2-16. *Describes table user*

Metadata has been changed, and columns (first_name and last_name) modified via Thrift are no longer available with CQL3! Don't worry! Data is not lost as CQL3 and Thrift rely on the same storage engine, and we can always get that metadata back by rebuilding them.

5. Let's rebuild first_name and last_name:

```
alter table user add first_name text;
alter table user add last_name text;
```

The problem is with CQL3's sparse tables. CQL3 has different metadata (CQL3Metadata) that has NOT been added to Thrift's CFMetaData. Do not mix and match CQL3 and Thrift to perform DDL/DML operations. It will always lead any one of these metadata to an inconsistent state.

A developer who can't afford loosing Thrift's dynamic column support still prefers to perform an insert via Thrift, but to read them back via CQL3. It is recommended to use CQL3 for a new application development over Cassandra. However, it has been noticed that Thrift based mutation still works faster than CQL3 (such as batch operation) until Cassandra 1.x.x releases. This is scheduled to address with Cassandra 2.0.0 release (https://issues.apache.org/jira/browse/CASSANDRA-4693).

Changing Data Types

Changing data types with Cassandra is possible in two ways, Thrift and CQL3.

Thrift Way

Let's discuss more about data types with legacy Thrift API:

1. Let's create a column family with minimal definition, such as:

```
create keyspace twitter with strategy_options={replication_factor:1} and
placement_strategy='org.apache.cassandra.locator.SimpleStrategy';

use twitter;
create column family default;
```

Default data type for comparator and validator is BytesType.

2. Let's describe the keyspace and have a look at the default column family (see Figure 2-17):

```
describe twitter;
```

```
Keyspace: twitter:
  Replication Strategy: org.apache.cassandra.locator.NetworkTopologyStrategy
  Durable Writes: true
    Options: [datacenter1:1]
  Column Families:
    ColumnFamily: default
      Key Validation Class: org.apache.cassandra.db.marshal.BytesType
      Default column value validator: org.apache.cassandra.db.marshal.BytesType
      Columns sorted by: org.apache.cassandra.db.marshal.BytesType
      GC grace seconds: 864000
      Compaction min/max thresholds: 4/32
      Read repair chance: 0.1
      DC Local Read repair chance: 0.0
      Populate IO Cache on flush: false
      Replicate on write: true
      Caching: KEYS_ONLY
      Bloom Filter FP chance: default
      Compaction Strategy: org.apache.cassandra.db.compaction.SizeTieredCompactionStrategy
      Compression Options:
        sstable_compression: org.apache.cassandra.io.compress.SnappyCompressor
```

Figure 2-17. *Structure of twitter keyspace*

3. Let's try to store some data in the column family:

```
set default[1]['type']='bytes'; gives an error
```

Figure 2-18 shows that this produces an error.

```
[default@twitter] set default[1]['type']='bytes';
org.apache.cassandra.db.marshal.MarshalException: cannot parse 'type' as hex bytes
```

Figure 2-18. *Error while storing string value but column value is of bytes type*

Since the comparator and validator are set to default data type (e.g., BytesType), Cassandra-cli is not able to parse and store such requests.

4. To get step 3 working, we need to use the assume function to provide some hint:

```
assume default keys as UTF8Type;
assume default comparator as UTF8Type;
assume default validator as UTF8Type;
```

5. Now let's try to change the comparator from BytesType to UTF8Type:

```
update column family default with comparator='UTF8Type';
gives error
```

This generates an error because changing the comparator type is not allowed (see Figure 2-19).

```
[default@twitter] update column family default with comparator='UTF8Type';
comparators do not match or are not compatible.
```

Figure 2-19. *Changing comparator type is not allowed*

6. Although changing comparator type is not allowed, we can always change the data type of the column and key validation class as follows:

```
update column family default with key_validation_class=UTF8Type and
default_validation_class = UTF8Type;
```

Columns in a row are sorted by column names and that's where comparator plays a vital role. Based on comparator type (i.e., UTF8Type, Int32Type, etc.) columns can be stored in a sorted manner.

CQL3 Way

Cassandra CQL3 is the driving factor at present. Most of the high-level APIs are supporting and extending further development around it.

Let's discuss a few tricks while dealing with data types in CQL3 way. We will explore with the default column family created in the Thrift way (see the preceding section).

1. Let's try to fetch rows from the default column family (see Figure 2-20).

```
Select * from default;
```

```
cqlsh:twitter> select * from default;

 key  | column1 | value
------+---------+---------------
 0x31 |    type | 0x6279746573
```

Figure 2-20. *Retrieving values using cql shell*

2. Let's issue the assume command and try to fetch rows from the default column family in readable format:

```
assume default(column1) values are text;
assume default(value) values are text;
assume default(key) values are text;
select * from default;
```

Figure 2-21 shows the result.

```
cqlsh:twitter> select * from default;

 key | column1 | value
-----+---------+-------
   1 |    type | bytes
```

Figure 2-21. *Retrieving after assume function is applied*

3. typeAsBlob or blobAsType functions can also be used to marshal data while running CQL3 queries:

```
select blobAsText(key),blobAsText(type),blobAsText(value) from default;
```

4. We can alter the data type of validator as follows:

```
alter table default alter value type text;
alter table default alter key type text;
```

■ **Note** The assume command will not be available after Cassandra 1.2.X release. As an alternative we can use typeAsBlob (e.g., textAsBlob) CQL3 functions.

Counter Column

Distributed counters are incremental values of a column partitioned across multiple nodes. Counter columns can be useful to provide counts and aggregation analytics for Cassandra-powered applications (e.g., Number of page hits, number of active users, etc.).

In Cassandra, a counter is a 64-bit signed integer. A write on counter will require a read from replica nodes (this depends on consistency level, default is ONE). While reading a counter column value, read has to be consistent.

Counter Column with and without replicate_on_write

Default value of replicate_on_write is true. If set to false it will replicate on one replica node (irrespective of replication factor). That might be helpful to avoid read-before-write on serving write request. But any subsequent read may not be consistent and may also result in data loss (single replica node is gone!).

Play with Counter Columns

In Chapter 1 we discussed setting multiple clusters on a single machine. First let's start with a cluster of three nodes on a single machine. (Please refer to the "Configuring Multiple Nodes on a Single Machine" section in Chapter 1.) In this recipe we will discuss the do's and don'ts of using counter columns.

1. Let's create a keyspace **counterkeyspace**:

```
create keyspace counterkeyspace with replication = {'class' : 'SimpleStrategy',
'replication_factor' : 2 }
```

2. Create a column family **counternoreptable** with replicate_on_write as false:

```
create table counternoreptable(id text PRIMARY KEY, pagecount counter)
with replicate_on_write='false';
```

3. Update pagecount to increment by 2 as follows:

```
update counternoreptable set pagecount=pagecount+2 where id = '1';
```

4. Select from the column family as follows:

```
select * from counternoreptable;
```

As shown in Figure 2-22, it results in zero rows. Whether it results in zero rows may depend on which node it is written to.

```
cqlsh:counterkeyspace> select * from counternoreptable;
cqlsh:counterkeyspace> █
```

Figure 2-22. *Inconsistent result on fetching from counter table*

5. Let's update pagecount for some more values and verify the results:

```
update counternoreptable set pagecount=pagecount+12 where id = '1';
select * from counternoreptable;
```

Figure 2-23 shows the result of this command.

```
update counternoreptable set pagecount=pagecount-2 where id = '1';
select * from counternoreptable;
```

```
cqlsh:counterkeyspace> select * from counternoreptable;

 id | pagecount
----+-----------
  1 |        12
```

Figure 2-23. *Retrieving from the counter table after incrementing the counter column value*

The result is different for this command (see Figure 2-24).

```
cqlsh:counterkeyspace> select * from counternoreptable;

 id | pagecount
----+-----------
  1 |        10
```

Figure 2-24. *Inconsistent result of counter column without replication*

You can see the inconsistent results on read with `replicate_on_write` as `false`. With this, conclude that by disabling such parameters we may avoid read-before-write on each write request, but subsequent read requests may result in inconsistent data. Also without replication we may suffer data loss, if a single replica containing an updated counter value goes down or is damaged. Try the above recipe with `replicate_on_write` as `true` and monitor whether results are consistent and accurate or not!

■ **Note**   You may refer to `https://issues.apache.org/jira/browse/CASSANDRA-1072` for more on counter columns.

Data Modeling Tips

Cassandra is a column-oriented database that is quite different from traditional RDBMS. We don't need to define schema up front, but it is always better to get a good understanding of the requirements and database before moving ahead with data modeling, including:

- Writes in Cassandra are relatively fast but reads are not. Pre-analysis of how we want to perform read operations is always very important to keep in mind before data modeling.

- Data should be de-normalized as much as possible.

- Choose the correct partitioning strategy to avoid rebuilding/populating data over updated partitioning strategy.

- Prefer using surrogate keys and composite keys (over super columns) while modeling table/column family.

Summary

To summarize a few things discussed in this chapter so far:

- Do not mix Thrift and CQL3 for DDL and DML operations, although reads should be fine.

- Avoid changing data types.

- Use Cassandra collection support for adding columns on the fly.

In Chapter 3, we will continue our discussion by exploring indexes, composite columns, and the latest features introduced in Cassandra 2.0, such as Compare and Set.

■ ■ ■

Indexes and Composite Columns

In previous chapters we have discussed big data problems, Cassandra data modeling concepts, and various schema management techniques. Although you should avoid normalizing the form of your data too much, you still need to model read requirements around columns rather than primary keys in your database applications.

The following topics will be covered in this chapter

- Indexing concept

- Data partitioning

- Cassandra read/write mechanism

- Secondary indexes

- Composite columns

- What's new in Cassandra 2.0

Indexes

An index in a database is a data structure for faster retrieval of a data set in a table. Indexes can be made over single or multiple columns.

Indexing is a process to create and manage a data structure called Index for fast data retrieval. Each index consists of indexed field values and references to physical records. In some cases a reference can be an actual row itself. We will discuss these cases in the **clustered indexes** section.

Physically data is stored on blocks in data structure form (like sstable in Cassandra). These data blocks are unordered and distributed across multiple nodes. Accessing data records without a primary key or index would require a linear search across multiple nodes. Let's discuss format index data structure.

Indexes are stored in sorted order into B-tree (balanced tree) structure, where indexes are leaf nodes under branch nodes. Figure 3-1 depicts data storage where multi-level leaf nodes (0,1) are indexed in sorted order and data is in unsorted order. Here each leaf node is a b-tree node containing multiple keys. Based on inserts/updates/deletes, the number of keys per b-tree node keeps changing but in sorted order.

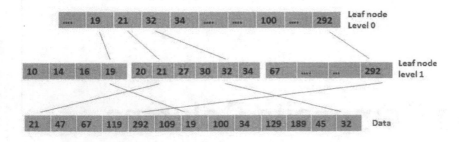

Figure 3-1. *b-tree Index and data structure with multi-level leaf nodes*

Let's simplify further. In Figure 3-2, the table containing age and row keys are leaf nodes and the other one is a physical table.

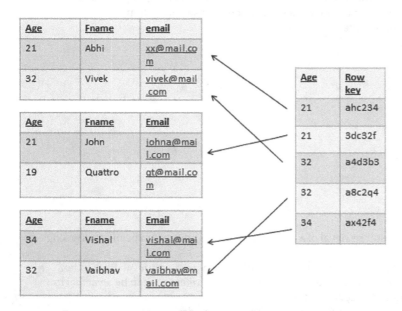

Figure 3-2. *A physical table and an index table as leaf node*

This allows faster retrieval of records using binary search. Since b-tree keeps data sorted for faster searching, it would introduce some overhead on insert, update, and delete operations and would require rearranging indexes. B-tree is the preferred data structure of a larger set of read and writes, that's why it's widely used with distributed databases.

Clustered Indexes vs. Non-Clustered Indexes

Indexes that are maintained independently from physical rows and don't manage ordering of rows are called non-clustered indexes (see Figure 3-1). On the other hand, clustered indexes will store actual rows in sorted order for the index field. Since a clustered index will store and manage ordering of physical rows, only one clustered index is possible per table.

The important question is for what scenarios we should use clustered indexes and non-clustered indexes. For example, a department can be multiple employees (many-to-one relation) and often is required to read employee details by department. Here department is a suitable candidate for a clustered index. All rows containing employee details would be stored and ordered by department for faster retrieval. Here employee name is a perfect candidate for a non-clustered index and thus we can hold multiple non-clustered indexes in a table but there will always be a single clustered index per table.

Index Distribution

With distributed databases, data gets distributed and replicated across multiple nodes. Retrieval of a data collection would require fetching rows from multiple nodes. Opting for indexes over a non-row key column would also require being distributed across multiple nodes, such as shards. Long-running queries can benefit from such shard-based indexing for fast retrieval of data sets.

Due to peer-to-peer architecture each node in a Cassandra cluster will hold an identical configuration. Data replication, eventual consistency, and partitioning schema are two important aspects of data distribution.

Please refer to Chapter 1 for more details about replication factor, strategy class, and read/write consistency.

Indexing in Cassandra

Data on a Cassandra node is stored locally for each row. Rows are distributed across multiple nodes, but all columns for a particular row key will be stored locally on a node. Cassandra by default provides the primary index over row key for faster retrieval by row key.

Secondary Indexes

Indexes over column values are known as secondary indexes. These indexes are stored locally on a node where physical data resides. That allows Cassandra to perform faster index-based retrieval of data. Secondary indexes are stored in a hidden column family and internally managed by the node itself.

Let's explore more on secondary indexes with a simple exercise.

1. First, let's create a keyspace **twitter** and column family **users**.

   ```
   create keyspace twitter with replication = { 'class':'SimpleStrategy' ,
   'replication_factor':3};use twitter;
   create column family users with key_validation_class='UTF8Type' and
   comparator='UTF8Type' and default_validation_class='UTF8Type';

   create table users (user_id uuid primary key, first_name text, twitter_handle text);
   ```

2. Let's create index over first_name using **create index** syntax (see Figure 3-3).

   ```
   create index fname_idx on users(first_name);
   ```

3. Describe table **users**:

   ```
   describe table users;
   ```

45

Figure 3-3 shows users schema with index created on first_name.

Index script

```
cqlsh:twitter> describe table users;

CREATE TABLE users (
    user_id timeuuid,
    first_name text,
    twitter_handle text,
    PRIMARY KEY (user_id)
) WITH
    bloom_filter_fp_chance=0.010000 AND
    caching='KEYS_ONLY' AND
    comment='' AND
    dclocal_read_repair_chance=0.000000 AND
    gc_grace_seconds=864000 AND
    index_interval=128 AND
    read_repair_chance=0.100000 AND
    replicate_on_write='true' AND
    populate_io_cache_on_flush='false' AND
    default_time_to_live=0 AND
    speculative_retry='99.0PERCENTILE' AND
    memtable_flush_period_in_ms=0 AND
    compaction={'class': 'SizeTieredCompactionStrategy'} AND
    compression={'sstable_compression': 'LZ4Compressor'};

CREATE INDEX fname_idx ON users (first_name);
```

Figure 3-3. *Users table with index on first_name*

4. Let's insert a few rows in the **users** column family.

   ```
   insert into users(user_id,first_name,twitter_handle) values(now(),'apress','#apress_team');
   insert into users(user_id,first_name,twitter_handle) values(now(),'jonathan','#jhassell');
   insert into users(user_id,first_name,twitter_handle) values(now(),'vivek','#mevivs');
   insert into users(user_id,first_name,twitter_handle) values(now(),'vivek','#vivekab');
   ```

5. Let's try to find records using the indexed column first_name (see Figure 3-4).

   ```
   select * from users where first_name='vivek';
   ```

Figure 3-4 shows output of fetching users having first name **vivek**.

```
cqlsh:twitter> select * from users where first_name='vivek';

 user_id                              | first_name | twitter_handle
--------------------------------------+------------+----------------
 498d9f00-d47f-11e3-be07-79e7dcea6dd7 |      vivek |        #mevivs
 64026320-d47f-11e3-be07-79e7dcea6dd7 |      vivek |       #vivekab

(2 rows)
```

Figure 3-4. *Fetching users by first_name*

Query over indexed column first_name with value 'vivek' (Figure 3-4) returns two rows. Here both rows can be available on the same node or different nodes.

One point worth mentioning here is that indexes would also be stored locally along with data rows, which would ensure data locality.

On the other hand, if we try to fetch rows using column twitter_handle, which is non-indexed:

```
select * from users where twitter_handle='#imvivek';
```

it results in the following error:

```
Bad Request: No indexed columns present in by-columns clause with Equal operator
```

Let's try another type of example:

1. Add another column **age**:

   ```
   alter table users  add age int  ;
   ```

2. Update a few rows for **age**:

   ```
   update users set age = 21 where user_id = 498d9f00-d47f-11e3-be07-79e7dcea6dd7;
   update users set age = 51 where user_id = 498b06f0-d47f-11e3-be07-79e7dcea6dd7;
   ```

3. Create an index on the **age** column:

   ```
   create index age_idx on users(age);
   ```

4. Now, try to fetch user_id for a user with age 21:

   ```
   select user_id from users where age =21;

    user_id
   --------------------------------------
    498d9f00-d47f-11e3-be07-79e7dcea6dd7

   (1 rows)
   ```

5. Next, let's try to fetch users whose age is greater than 21:

   ```
   select user_id from users where age >=21;
   ```

It results in the same error we saw in the preceding example:

```
Bad Request: No indexed columns present in by-columns clause with Equal operator
```

Secondary indexes allow users to retrieve records using indexed columns with "=" only. But still real-time applications would want to perform range queries over non-row key columns.

Columns with low cardinality values are generally recommended for indexing as probability of retrieving data from various nodes in a cluster is very high. Whereas a column with high cardinality may result in performance overhead under high data load and large Cassandra clusters. Indexes on unique columns should be fine in case data volume is not huge. Also, you can turn on tracing to monitor queries on secondary indexes. In general, tracing can be used for investigating CQL3 queries. You can turn on/off tracing as

```
TRACING ON;
TRACING OFF;
```

From version 1.2 onward Cassandra provides composite column support. Using composite column support it is possible to perform range scan over composite columns.

Let's explore composite columns more in the next section.

Composite Columns

Thinner rows distributed across multiple nodes in a cluster work fine with added secondary index support. Still applications may require applying sorting order on fetched rows. Implementing such sorting mechanisms at the client side with a massive amount of data is clearly not desirable. Applications would rely on a database for such supports.

In version 1.1 onward, Cassandra provides support for sorted wide rows via composite columns. Sorted wide rows mean more columns in place of skinny rows, which allows more data to be stored in a colocated way.

Let's explore it with the same Twitter example, where the user and their tweets should be stored locally and need to be sorted by tweet_date.

1. First, let's create a keyspace **twitter** and column family **users**:

```
create keyspace twitter with replication = {'class':'SimpleStrategy',
'replication_factor':1};
use twitter;

create table users(user_id text,followers set<text>, tweet_date timestamp,
tweet_body text, first_name text, PRIMARY KEY(user_id,tweet_date, first_name));
```

Here for **users** the table contains a composite primary key with user_id, tweet_date and first_name.

2. Let's store a few columns in the **users** column family for user_id **'imvivek'**:

```
// insert records for 'imvivek'
 insert into users(user_id,tweet_date,tweet_body,first_name)
values('imvivek','2013-12-31','good bye 2013','vivek');
 insert into users(user_id,tweet_date,tweet_body,first_name)
values('imvivek','2014-01-01','welcome 2014','vivek');
 insert into users(user_id,tweet_date,tweet_body,first_name)
values('imvivek','2014-01-04','Working on Cassandra book on weekend','vivek');
```

3. Let's store a few columns in the **users** column family for user_id **'jhassell'**:

```
// insert records for 'jhassel'
 insert into users(user_id,tweet_date,tweet_body,first_name)
values('jhassell','2013-12-21','2013 bye bye','jonathan');
 insert into users(user_id,tweet_date,tweet_body,first_name)
values('jhassell','2014-01-01','2014! another exciting year','jonathan');
 insert into users(user_id,tweet_date,tweet_body,first_name)
values('jhassell','2014-03-25','Cassandra book, rolling out!','jonathan');
```

4. Let's try to fetch rows ordered by tweet_date in descending order (see Figure 3-5).

```
select * from users where user_id='imvivek' order by tweet_date DESC;
```

Figure 3-5 output shows rows sorted by tweet_date for user_id imvivek

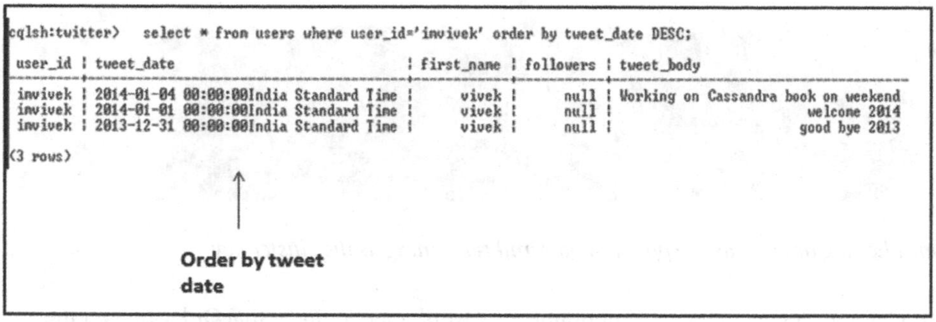

Figure 3-5. Users with user_id imvivek, sorted by tweet date

5. We can also query the **users** table to fetch rows for multiple user ids sorted by tweet date as follows (see Figure 3-6):

```
select * from users where user_id in('imvivek','jhassell') order by tweet_date DESC;
```

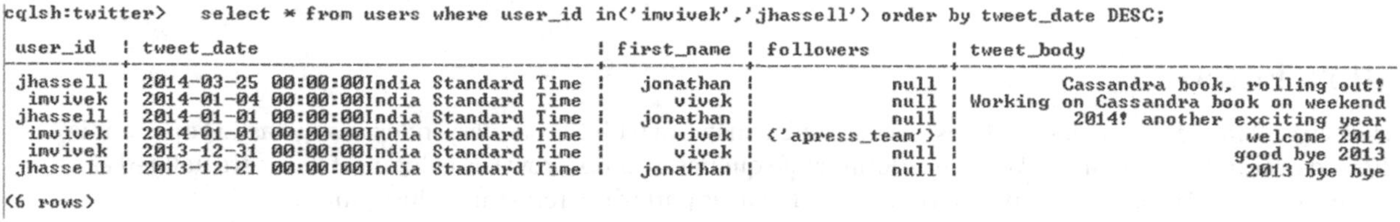

Figure 3-6. Here we are retrieving rows for user_id imvivek and jhassel using the in clause

Figure 3-6 shows rows sorted by tweet_date for users imvivek and jhassel.

6. Let's update a few rows for **followers** and explore how the data structure would look internally:

```
update users set followers = {'jhassell'} where user_id='imvivek' and
tweet_date = '2013-12-31' and first_name = 'vivek';

update users set followers = {'jhassell'} where user_id='imvivek' and
tweet_date = '2014-01-01' and first_name = 'vivek';
```

Here, the user_id will be a partition key and the rest of the columns tweet_date and first_name will be part of a clustering key. Let's try to look on stored data with the help of Figure 3-7.

Figure 3-7. Composite column having imvivek as the partition key and remaining as the cluster key

While performing data modeling over Cassandra, one important point worth mentioning is that Ordering over the clustering key works only if the partition key is in EQ/IN clause. Hence, while defining table structure, make sure that columns that need to be sorted are part of the cluster columns. You can also ensure clustering order while creating table as

```
create table users(user_id text,followers set<text>, tweet_date timestamp, tweet_body text,
first_name text, PRIMARY KEY(user_id,tweet_date, first_name)) with clustering order
by (tweet_date DESC);
```

This would ensure keeping clustered data ordered by tweet_date in descending order.

Allow Filtering

As discussed in the "Secondary Indexes" section, one constraint with a secondary index is only the EQ operator is allowed over indexed columns. We can perform range queries over non-row key columns if those are defined as cluster columns. The query below will work as tweet_date is part of the remaining clustering key.

By default Cassandra restricts execution of such queries and intelligently educates them about performance overhead of such queries by explicitly asking to enable filtering using **allow filtering** clause.

Users can run such queries by explicitly adding the ALLOW FILTERING clause with select queries as

```
select * from users where tweet_date>='2013-12-31' allow filtering;
```

Figure 3-8 shows the output from selecting users by tweet date with allow filtering enabled.

```
cqlsh:twitter> select * from users where tweet_date>='2013-12-31' allow filtering;

 user_id  | tweet_date                       | first_name | followers        | tweet_body
----------+----------------------------------+------------+------------------+----------------------------------------
  imvivek | 2013-12-31 00:00:00India Standard Time |      vivek | {'jhassell'}     |                          good bye 2013
  imvivek | 2014-01-01 00:00:00India Standard Time |      vivek | {'apress_team'}  |                           welcome 2014
  imvivek | 2014-01-04 00:00:00India Standard Time |      vivek |            null  | Working on Cassandra book on weekend
 jhassell | 2014-01-01 00:00:00India Standard Time |   jonathan |            null  |         2014! another exciting year
 jhassell | 2014-03-25 00:00:00India Standard Time |   jonathan |            null  |         Cassandra book, rolling out!

(5 rows)
```

Figure 3-8. Rows fetched for tweet date greater than or equal to 31st December 2013 with the filtering allowed option

With a very large amount of data being distributed within a large cluster, running such queries may degrade the application's performance, if a partition key is not present with EQ/IN clause. It is recommended that users analyze data first before running such queries, rather than perform a range of queries specific to a partition key.

Expiring Columns

We can also set the TTL (Time To Live) value per each column in a Cassandra table. Upon exceding the defined TTL, the column would be logically deleted and marked as obsolete. Though memory will only be freed up during compaction.

For example, with the banking system a OTP (One Time Password) is generated per request, and would expire after a certain time. For such scenarios, we can temporarily create a column with TTL value.

Prior to Cassandra 2.x, setting the TTL value was possible via insert and update operations. Now with 2.x we can also set the default TTL value for per column family/table. For storing TTL and expiration time, the column family would required an additional eight bytes.

Let's explore TTL with a simple exercise.

1. First, let's create a keyspace **demo** and column family **TTLSample**

   ```
   create keyspace demo with replication = {'class':'SimpleStrategy',
   'replication_factor':1};
   use demo;

   create table "TTLsample"(user_id text PRIMARY KEY, OTP text, name text);
   ```

2. Let's insert a record in **TTLsample** table with TTL value set to 16 seconds

   ```
   insert into "TTLsample"(user_id,OTP,name) values('imvivek','12121','vivek') using TTL 16;
   ```

Upon running the above query a row will be stored in **TTLsample** table with TTL value set to 16 seconds for each column.

3. Let's try to fetch the row before and after 16 seconds

   ```
   select * from "TTLsample";
   ```

Figure 3-9 shows the output before and after 16 seconds.

```
                    cqlsh:demo> select * from "TTLsample";
Before TTL expiry ──→   user_id | name  | otp
                    ---------+--------+--------
                     imvivek | vivek | 12121

                    (1 rows)

                    cqlsh:demo> select * from "TTLsample";
After TTL expiry ──→
                    (0 rows)
```

Figure 3-9. *Output before and after the time to live expiration*

4. You also can set a TTL value for each column with an insert or update operation.

   ```
   insert into "TTLsample"(user_id,name) values('imvivek','vivek');
   insert into "TTLsample"(user_id,OTP) values('imvivek','12121') using TTL 16;
   ```

Figure 3-10 shows output of the otp column before and after TTL expiration.

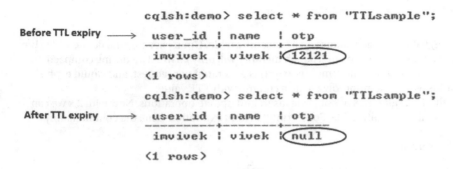

Figure 3-10. *Before and after TTL expiry for OTP column*

5. It is also possible to reset the TTL value before its expiration

```
insert into "TTLsample"(user_id,OTP) values('imvivek','12121') using TTL 160;
// expires OTP after 160 seconds

 insert into "TTLsample"(user_id,OTP) values('imvivek','12121') using TTL 16;
// reset expiry to 16 seonds
```

Default TTL

With Cassandra 2.x, it is also possible to set the default TTL per column family:

```
alter table "TTLsample" with default_time_to_live=10;
```

Figure 3-11 shows the default TTL value set to 10 seconds.

```
cqlsh:demo> describe table "TTLsample";

CREATE TABLE "TTLsample" (
  user_id text,
  name text,
  otp text,
  PRIMARY KEY (user_id)
) WITH
  bloom_filter_fp_chance=0.010000 AND
  caching='KEYS_ONLY' AND
  comment='' AND
  dclocal_read_repair_chance=0.000000 AND
  gc_grace_seconds=864000 AND
  index_interval=128 AND
  read_repair_chance=0.100000 AND
  replicate_on_write='true' AND
  populate_io_cache_on_flush='false' AND
  default_time_to_live=10 AND
  speculative_retry='99.0PERCENTILE' AND
  memtable_flush_period_in_ms=0 AND
  compaction={'class': 'SizeTieredCompactionStrategy'} AND
  compression={'sstable_compression': 'LZ4Compressor'};
```

Default TTL ⟶ (points to `default_time_to_live=10`)

Figure 3-11. *Output shows default TTL set to 10 seconds*

Data Partitioning

Partitioning schema plays an important role in data distribution across nodes. Cassandra offers three types of data partitioning strategies:

RandomPartitioner. Selecting the Random partitioner would distribute data across the nodes using MD5 hashing algorithm. The random partitioner would evenly distribute data in a cluster. Data distribution would rely on assigned initial_token value or num_tokens for assigning rows to each node. MD5 hashes are 16 bytes and are used to represent hexadecimal digits. Each node is assigned a data range that is represented by the token value. On receiving read/write request, with random partitioner selected as the partitioning strategy hash value for each row key gets generated and assigned to the node responsible for serving that read/write request. That's how data gets distributed with random partitioning.

Murmur3Partitioner. Default partitioner. Similar to RandomPartitioner but uses the Murmur Hash function for calculating the token value. Murmur3 hash represents a 32- or 128-bit value, with Cassandra it uses a 128-bit value for tokens. Another difference with MD5 and murmur3 hashes is that the latter is noncryptographic hashing whereas MD5 is cryptographic hashing function (e.g., one-way hashing).

ByteOrderedPartitioner. Recommended for sequential data access. With this partitioner, data gets distributed across the nodes according to row key values in a cluster. The frequency of uneven data distribution is high and may lead to performance hotspots in case of frequent writes within a specific range.

Changing Partitioners

Decide on a partitioning strategy up front while doing data modeling. Changing partitioners while a cluster is already accepting data is not allowed and Cassandra will throw an error such as:

```
ERROR 23:34:10,630 Cannot open \var\lib\cassandra\data\system\schema_keyspaces\system-schema_keyspaces-jb-1; partitioner org.apache.cassandra.dht.Murmur3Partitioner does not match system partitioner org.apache.cassandra.dht.RandomPartitioner.
```

Note that the default partitioner starting with Cassandra 1.2 is Murmur3Partitioner, so you will need to edit that to match your old partitioner if upgrading.

Data Colocation

Columns for each row key would be stored locally on that node. Based on the replication factor and strategy, data would be replicated across the nodes. That's where Cassandra ensures data locality and high availability.

Let's discuss how reads and writes work in Cassandra.

Cassandra Writes

With peer-to-peer architecture every node in the cluster is eligible to receive read/write requests. Nodes that work as a proxy or delegator between the client and data node (assigned to serve the write requests) are called Coordinator nodes. Upon receiving the write request, columns would first be written on the commit log and then into the memtable onto assigned node. Based on supplied consistency, an acknowledgement would be sent back to the client.

Data would be flushed out in the form of **sstable** on to a disk based on **memtable_total_space_in_mb**. Upon exceeding the limit, the largest memtable will be flushed out on to disk. Figure 3-12 shows the image representations for Cassandra inserts and updates.

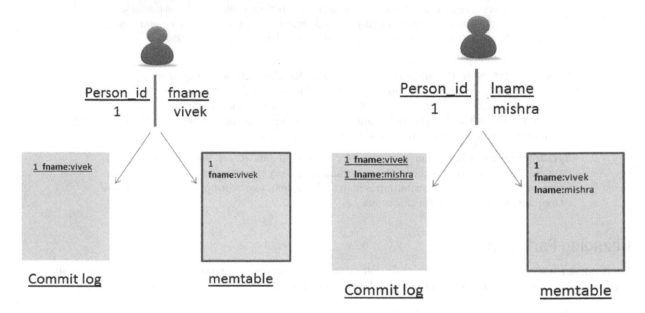

Figure 3-12. *a) The image on the left, insert fname for id 1; b) the image on the right, update to add column lname for person id 1*

In Figure 3-12a and 3-12b, columns fname and lname are stored in successive write column requests. Each memtable is an in-memory representation of a column family. Whereas the commit log will keep data in the same sequence. Inserting a record for Person column family would keep adding it to corresponding memtable. For example, adding another column age for row key 1 and updating fname, would modify the data structure as shown in Figure 3-13.

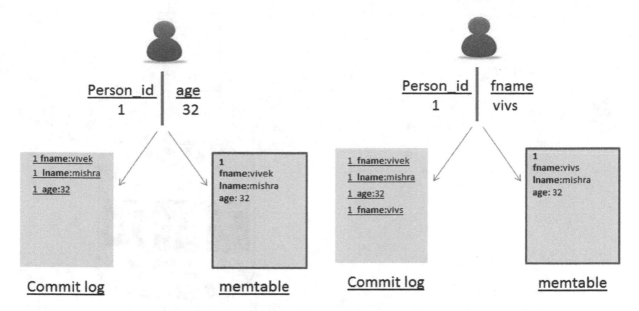

Figure 3-13. *a) The image on the left shows adding the age column with value 32; b) the one on the right shows updating the row for person id 1 with column fname for value vivs*

Figure 3-13a and 3-13b depicts the insert of age column and modifying **fname** respectively. Please note that Cassandra provides faster write throughput, because there are no updates but only insert operations. But based on the timestamp, the row with the latest timestamp will be returned as output. Here the timestamp is internally managed by Cassandra itself. The process of compaction will manage freeing space by deletion of obsolete or tombstone rows. With version 2.x Cassandra provides Compare and Set (CAS) support. We will discuss this later in the "Compare and Set" section.

Cassandra Reads

The process of coordinator selection and assigning a read request to data node is similar to a write request. Cassandra returns columns for a particular row key with the latest timestamp. Cassandra internally performs the following steps to return the column name and value:

1. First search memtable for cached values

2. Scan sstables using bloom filter row key for column

3. Sort eligible sstables by latest timestamp

4. Merge and return columns with recent timestamp value

We will discuss bloom filter in detail in Chapter 8.

Figure 3-14 shows how read works with Cassandra storage architecture.

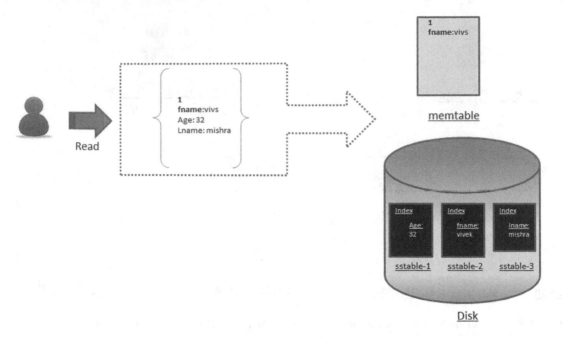

Figure 3-14. *Read mechanism with Cassandra*

What's New in Cassandra 2.0

Cassandra 2.0 and subsequent versions came with lots of new features and many performance enhancements on the CQL3 side. Cassandra 2.0 introduces support for several key capabilities related to indexes and composite columns, including:

- Compare and set (CAS)
- Secondary indexes over composite columns
- Conditional DDL

■ **Note**   Before this book went to press, Cassandra 2.1 was released. Please see Chapter 11 for information on several important new features in version 2.1.

Compare and Set

With eventual consistency and replication support, transactional data can be consistent and highly available. But still in a few cases, applications might need to access data independently across multiple threads, such as an application need to generate unique OTP (One Time Password) and want to avoid race conditions. Support for compare and set (CAS) has been enabled since 2.x onwards.

■ **Note**   CAS support in Cassandra is based on paxos algorithm. You may refer to
http://en.wikipedia.org/wiki/Paxos_(computer_science) for more details.

Algorithm

A compare and set operation requires a read of the latest value from the replica node, and the node acting as a coordinator is called the Leader. The Leader will coordinate the replica nodes so they perform the following tasks:

1. Prepare token and promise

2. Issue Read to replica and receive output with the latest token

3. Propose updated value and get acceptance

4. Issue commit to each replica and receive acknowledgement

Using CAS

Each CAS operation will require multiple DB round trips; hence, it is recommended to use such operations for limited requirements to avoid performance overhead.

Let's use CAS with the same Twitter example.

1. First, let's create a keyspace **twitter** and column family **users**

```
create keyspace twitter with replication={'class':'SimpleStrategy','replication_factor': 1 };
 use twitter;
create table users(user_id text primary key,followers set<text>, tweet_date timestamp,
tweet_body text, first_name text);
```

2. Let's insert a record for row key 'imvivek' using the if not exists command

```
insert into users(user_id,tweet_date,tweet_body,first_name) values('imvivek','2013-12-31',
'good bye 2013','vivek')  if not exists;
```

3. Let's re-run the same command with first_name as 'vivek_update'

```
insert into users(user_id,tweet_date,tweet_body,first_name) values('imvivek','2013-12-31',
'good bye 2013','vivek_update')  if not exists;
```

In Figure 3-15, since before executing the command at step 2 the table was empty and no records existed; hence, it gets successfully applied, but the same fails at step 3.

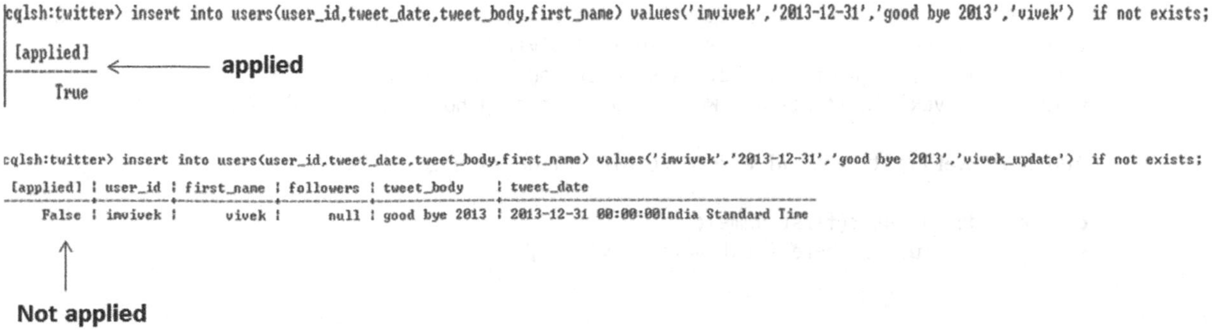

Figure 3-15. A CAS operation to insert a row with id imvivek if it doesn't exist

4. Let's try to use CAS to update users for adding followers for a user having the first name "vivek"

```
update users set followers = {'apress'} where user_id='imvivek' if first_name='vivek'
```

```
select * from users;
```

Figure 3-16 shows the result of retrieving records after the CAS update operation.

```
cqlsh:twitter> select * from users;

 user_id | first_name | followers  | tweet_body    | tweet_date
---------+------------+------------+---------------+--------------------------------------------
 imvivek |      vivek | {'apress'} | good bye 2013 | 2013-12-31 00:00:00India Standard Time

(1 rows)
```

Figure 3-16. *Rows after succesfully updating using the CAS operation*

Secondary Index over Composite Columns

With Cassandra 2.x onward, Cassandra allows the secondary index over clustering keys, which allows the user to run queries using EQ operater. Prior to 2.x, it was possible to include the remaining cluster key in a where clause only if the preceding clustering key part was available in the where clause.

Let's use the same Twitter example discussed within the "Composite Columns" section.

1. First, let's create a keyspace **twitter** and column family **users**:

```
create keyspace twitter with replication = {'class':'SimpleStrategy',
'replication_factor':1};
use twitter;

create table users(user_id text,followers set<text>, tweet_date timestamp,
tweet_body text, first_name text, PRIMARY KEY(user_id,tweet_date, first_name));
```

2. Let's Store a few colums in the **users** column family for user_id 'imvivek':

```
// insert records for 'imvivek'
 insert into users(user_id,tweet_date,tweet_body,first_name)
values('imvivek','2013-12-31','good bye 2013','vivek');
 insert into users(user_id,tweet_date,tweet_body,first_name)
values('imvivek','2014-01-01','welcome 2014','vivek');
 insert into users(user_id,tweet_date,tweet_body,first_name)
values('imvivek','2014-01-04','Working on Cassandra book on weekend','Vivs');
```

3. Add index over first_name and retrieve records with the first name **vivek**:

```
create index on user(first_name);
select * from users where first_name = 'vivek';
```

Figure 3-17 shows the result of fetching data by clustering the column first_name after secondary indexes enabled.

```
cqlsh:twitter> select * from users where first_name = 'vivek';

 user_id | tweet_date                                  | first_name | followers | tweet_body
---------+---------------------------------------------+------------+-----------+----------------
 imvivek | 2013-12-31 00:00:00India Standard Time      |    vivek   |     null  | good bye 2013
 imvivek | 2014-01-01 00:00:00India Standard Time      |    vivek   |     null  | welcome 2014

(2 rows)
```

Figure 3-17. *Fetch records after enabling secondary indexes over clustering the column first_name*

Conditional DDL

With Cassandra 2.x it is now possible to perform conditional DDL operations over keyspace, table, and indexes. A conditional DDL allows a user to validate whether a questioned keyspace, column family, or index is present or not. Let's look at a few examples of how this works.

Keyspaces

- Create the keyspace twitter if it doesn't exist:

  ```
  create keyspace if not exists twitter with replication = {'class':'SimpleStrategy',
  'replication_factor'
                        : 3};
  ```

- Drop the keyspace twitter if it exists:

  ```
  drop keyspace if exists twitter;
  ```

Tables

- Create a table **users** if it doesn't exist:

  ```
  create table if not exists users(user_id text,followers set<text>, tweet_date timestamp,
      tweet_body text, first_name text, PRIMARY KEY(user_id,tweet_date, first_name));
  ```

- Drop the table **users** if it exists:

  ```
  drop table if exists users;
  ```

Indexes

- Create an index over column first_name on the **users** table if it doesn't exist:

  ```
  create index if not exists users_first_name_idx on users(first_name);
  ```

- Drop the index users_first_name_idx if it exists:

  ```
  drop index if exists users_first_name_idx;
  ```

Summary

With this chapter we have discussed data modeling and indexing concepts and their use in Cassandra. To summarize, a few points are worth reiterating:

- Indexes are stored locally to data node.

- Updates to column values are indexes, which is an atomic operation.

- Avoid secondary indexes over high cardinality column values.

- Too many indexes can kill write performance and should be avoided. Instead try to denormalize the data model, or use composite columns.

With data modeling in place, the next question while designing an application with Cassandra that naturally comes to mind is data security. Chapter 4 will talk about data encryption/decryption and the security options available with Cassandra.

CHAPTER 4

■ ■ ■

Cassandra Data Security

Database security means protecting sensitive data and database applications from unauthorized access. Scalability and high availability of data are definitely good things to have, but so is data security! For example, financial and other public industries require scalable, highly available, and secured databases. That said, organizations of many kinds, and especially financial and governmental ones, prioritize security.

Since Cassandra 1.2.2 onward, we can secure data with Cassandra in two ways, access control and encryption. Authorization and authentication support is available with Cassandra.

This chapter will cover the following topics:

- Cassandra's system and system_auth keyspaces

- Managing user permissions

- Internode and client-server SSL encryption

Authentication and Authorization

Authentication means providing control over users trying to access a data store. The user's identity has to be validated while securing the connection with the database. There are three basic types of authentication:

- *Internal authentication.* With internal authentication generally the data store would manage the user's login via credentials such as user id and password.

- *External authentication.* External authentications, such as Kerberos, are network protocols to authenticate a client's identity using a ticketing system.

- *Client-server encryption.* Client-to-server or node-to-node encryption is another way by which data can be transferred across the cluster. With encryption, clients' or nodes' public trust certification has to be installed on another node.

The authentication process is limited to user verification and identification.

Processing user access control is known as authorization. Database authorization means managing a user's role and privileges to schema, tables, and columns.

Cassandra's internal authentication is an SSL-encryption mechanism that we'll look at in the form of practical recipes.

system and system_auth Keyspaces

The **system** keyspace contains information about available column families, columns, and clusters. The **system_auth** keyspace mainly contains authentication information, user credentials, and permissions. We will discuss them in upcoming recipes. Table/column families under the system keyspace are:

- schema_keyspace

- schema_columns

- schema_columnfamilies

- local

- peers

Figure 4-1 shows the system keyspace hierarchy, where the root node is the system keyspace, subordinate nodes are column families, and each table represents columns defined with those column families.

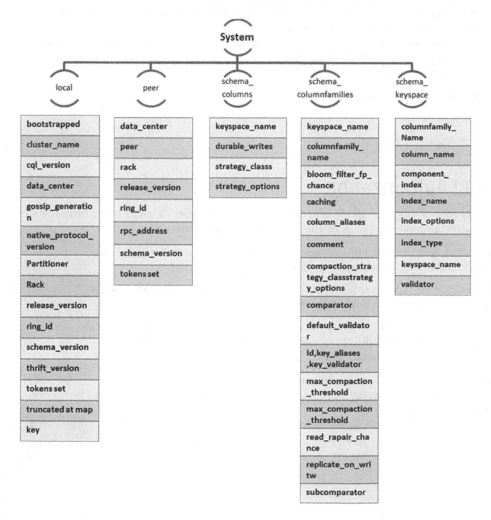

Figure 4-1. *System keyspace and underlying column families and corresponding columns*

Before we start managing authentication and authorization with system and system_auth keyspaces, let's start this section by exploring whether the system keyspace is modifiable!

The system Keyspace Is Unmodifiable

Before we jump into authorization and authentication, let's play with the system keyspace to validate whether it is modifiable by an external user. Since it is a keyspace internally managed by Cassandra's storage architecture, it is not possible for an external user to update/modify the system keyspace. The next simple exercise is all about trying these negative scenarios to validate this. While designing schema it is not required to tweak such exercise with Cassandra, but since we are talking about security, let's see how secure the system keyspace is!

Read access to the system keyspace is provided by default. Let's try accessing it with some exercises LocalStrategy.

The system keyspace is configured with LocalStrategy as replication options.

- First, let's try creating a keyspace with a replication strategy as LocalStrategy:

```
create keyspace system_test with replication = { 'class':'localstrategy',
'replication_factor':3};
create keyspace system_test with replication = { 'class':'LocalStrategy'};
```

Figure 4-2 shows that we cannot create another keyspace with LocalStrategy.

```
cqlsh:system_test> create keyspace system_test with replication = { 'class':'LocalStrategy','replication_factor':3};
Bad Request: Unrecognized strategy option {replication_factor} passed to LocalStrategy for keyspace system_test
text could not be lexed at line 1, char 48
cqlsh:system_test> create keyspace system_test with replication = { 'class':'LocalStrategy'};
Bad Request: Unable to use given strategy class: LocalStrategy is reserved for internal use.
text could not be lexed at line 1, char 48
```

Figure 4-2. *LocalStrategy is not for public use*

- Now, let's try to modify/drop the system keyspace:

```
alter keyspace system with replication = { 'class' : 'LocalStrategy'};alter keyspace
system with replication = { 'class' : 'SimpleStrategy', 'replication_factor':3};
```

Figure 4-3 shows it is not possible to alter the system keyspace for changing strategy.

```
cqlsh:system> alter keyspace system with replication = { 'class' : 'SimpleStrategy', 'replication_factor':3};
Bad Request: Cannot alter system keyspace
text could not be lexed at line 1, char 42
cqlsh:system> alter keyspace system with replication = { 'class' : 'LocalStrategy'};
Bad Request: Cannot alter system keyspace
text could not be lexed at line 1, char 42
```

Figure 4-3. *Changing the strategy class is not possible with the system keyspace*

This means we cannot alter the system keyspace either.

- Next, let's try to drop the system keyspace:

    drop keyspace system;

Figure 4-4 shows we cannot drop the system keyspace either!

```
cqlsh:system> drop keyspace system;
Bad Request: system keyspace is not user-modifiable
```

Figure 4-4. *Dropping system keyspace is not an option*

With these examples we can conclude that modifying the system keyspace is not permissible; however we can modify data present in column families under the system keyspace.

Accessing system_auth Keyspace with Authentication Enabled

Cassandra's binary distribution comes with two authenticators: PasswordAuthenticator and AllowAllAuthenticator. With PasswordAuthenticator enabled, Cassandra validates users by their names and hashed passwords. These user credentials are stored in the system_auth.credentials table.

In this recipe, we will discuss managing user credentials and accessing the system_auth keyspace using PasswordAuthenticator. With authentication enabled we need to provide the user name and password while connecting to the Cassandra server. Cassandra stores the user name and hashed password under the credentials table in this keyspace.

Before we start, we need to enable authentication in cassandra.yaml as follows:

authenticator: org.apache.cassandra.auth.PasswordAuthenticator

Now that authentication is enabled, you can complete the following steps:

1. First, let's try connecting to the server creating a keyspace without our username/password:

    create keyspace system_temp with replication = { 'class' : 'SimpleStrategy', 'replication_factor' : 3 };

On executing, the above command will produce the error shown in Figure 4-5.

```
[impadmin@impetus-NL052 kundera-neo4j]$ /home/impadmin/software/apache-cassandra-1.2.4/bin/cqlsh -3
Connected to Test Cluster at 127.0.0.1:9160.
[cqlsh 2.3.0 | Cassandra 0.0.0 | CQL spec 3.0.0 | Thrift protocol 19.35.0]
Use HELP for help.
cqlsh> create keyspace system_temp with replication = { 'class' : 'SimpleStrategy', 'replication_factor' : 3 };
Bad Request: You have not logged in
```

Figure 4-5. *A connection failure because we did not authenticate before trying to create the keyspace*

2.   Now, connect using the user name and password and create the keyspace:

```
// connect with user name and password.
$CASSANDRA_HOME/bin/cqlsh -3 -u cassandra -p cassandra

// create keyspace.
 create keyspace system_temp with replication = { 'class' : 'SimpleStrategy',
'replication_factor' : 3 };
```

Figure 4-6 shows a successful connection.

```
[impadmin@impetus-NL052 kundera-neo4j]$ /home/impadmin/software/apache-cassandra-1.2.4/bin/cqlsh -3 -u cassandra -p cassandra
Connected to Test Cluster at 127.0.0.1:9160.
[cqlsh 2.3.0 | Cassandra 1.2.4 | CQL spec 3.0.0 | Thrift protocol 19.35.0]
Use HELP for help.
cqlsh> create keyspace system_temp with replication = { 'class' : 'SimpleStrategy', 'replication_factor' : 3 };
cqlsh> 
```

Figure 4-6. *After successfully connecting using user credentials*

3.   Next, let's create a superuser and non-superuser:

```
create user normaluser with password 'normal';
create user 'superuser' with password 'superuser' superuser;
```

It will store user information in **users** table as shown in Figure 4-7.

```
cqlsh> use system_auth;
cqlsh:system_auth> select * from users;

 name        | super
-------------+-------
  cassandra  | True
    packtsu  | True
  superuser  | True
 normaluser  | False
```

Figure 4-7. *List of registered users*

Also, you can explore the credentials table for its password and other information.

With this we conclude that we can create and manage user credentials using authentication.

In the preceding recipe we managed to create a **superuser** and **non-superuser**. The next recipe will talk about managing and accessing user credentials on different column families for both types of users.

Managing User Permissions

Cassandra provides the mechanism to manage user permission and credentials for authenticated users. In this recipe, we will explore more about managing/accessing user permissions. The default authorizer set configured in cassandra.yaml is AllowAllAuthorizer.

1. First, let's try to authenticate with normaluser and manage its permission with AllowAllAuthorizer:

```
//connect with cqlsh.
$CASSANDRA_HOME/bin/cqlsh -3 -u normaluser -p normal

// list user permissions.
list all permissions of normaluser;
list all permissions of cassandra;

// grant permissions to normaluser.
grant select on all keyspaces to normaluser;
```

This results in the authorization error shown in Figure 4-8. Hence, we need to enable the authorizer in cassandra.yaml.

```
cqlsh> list all permissions of normaluser;
Bad Request: LIST PERMISSIONS operation is not supported by AllowAllAuthorizer
cqlsh> list all permissions of cassandra;
Bad Request: LIST PERMISSIONS operation is not supported by AllowAllAuthorizer
cqlsh> grant select on all keyspaces TO normaluser;
Bad Request: GRANT operation is not supported by AllowAllAuthorizer
```

Figure 4-8. *Managing permissions with AllowAllAuthorizer is not allowed*

2. Let's enable the authorizer configuration in cassandra.yaml and restart the server. Figure 4-9 shows how to enable CassandraAuthorizer in cassandra.yaml.

```
# - AllowAllAuthorizer allows any action to any user - set it to disable authorization.
# - CassandraAuthorizer stores permissions in system_auth.permissions table. Please
#   increase system_auth keyspace replication factor if you use this authorizer.
# authorizer: org.apache.cassandra.auth.AllowAllAuthorizer
authorizer: org.apache.cassandra.auth.CassandraAuthorizer
```

Figure 4-9. *Enabling CassandraAuthorizer in cassandra.yaml*

3. Now, let's try to create a keyspace with normaluser:

```
create keyspace testkeyspace with replication = { 'class' : 'SimpleStrategy',
'replication_factor' : 3};
```

Again, we get an error (see Figure 4-10) because create permissions have not been given to normaluser.

```
cqlsh> create keyspace testkeyspace with replication = { 'class' : 'SimpleStrategy' , 'replication_factor' : 3};
Bad Request: User normaluser has no CREATE permission on <all keyspaces> or any of its parents
```

Figure 4-10. *Error as normaluser doesn't have permission to create a keyspace*

Also, normaluser can't grant permissions to itself. Only superuser can!

4. So, let's log in with superuser:

```
$CASSANDRA_HOME/bin/cqlsh -3 -u superuser -p superuser
```

5. Now, create a keyspace and grant permissions to normaluser:

```
// create keyspace.
create keyspace testkeyspace with replication = { 'class' : 'SimpleStrategy' ,
'replication_factor' : 3};

// grant all permissions on 'testkeyspace' to normaluser.
grant all on keyspace testkeyspace to normaluser;
```

Upon issuing permission to normaluser, the server will store its permissions in the permissions table under the system_auth keyspace.

6. Let's explore all given permissions of normaluser (see Figure 4-11).

```
// list all permissions.
list all permissions of normaluser;
```

```
cqlsh> list all permissions of normaluser;

 username   | resource                    | permission
------------+-----------------------------+-----------
 normaluser | <keyspace testkeyspace>     |     CREATE
 normaluser | <keyspace testkeyspace>     |      ALTER
 normaluser | <keyspace testkeyspace>     |       DROP
 normaluser | <keyspace testkeyspace>     |     SELECT
 normaluser | <keyspace testkeyspace>     |     MODIFY
 normaluser | <keyspace testkeyspace>     |  AUTHORIZE
```

Figure 4-11. The current normaluser permissions

7. Let's create **anotherkeyspace** and a privileges table under it with the name **privileges**:

```
// create keyspace.
create keyspace anotherkeyspace with replication = { 'class' : 'SimpleStrategy' ,
'replication_factor' : 3};

// create table.
create table privileges (user_id text PRIMARY KEY, read_write boolean);
```

8. Let's grant all permissions on the **privileges** table to normaluser:

```
grant all on table anotherkeyspace.privileges to normaluser;
```

9. Now, let's list all `normaluser` permissions again (see Figure 4-12).

```
// list all permissions.
list all permissions of normaluser;
```

```
 username   | resource                              | permission
------------+---------------------------------------+------------
 normaluser | <table anotherkeyspace.privileges>    |     CREATE
 normaluser | <table anotherkeyspace.privileges>    |      ALTER
 normaluser | <table anotherkeyspace.privileges>    |       DROP
 normaluser | <table anotherkeyspace.privileges>    |     SELECT
 normaluser | <table anotherkeyspace.privileges>    |     MODIFY
 normaluser | <table anotherkeyspace.privileges>    |  AUTHORIZE
 normaluser |                <keyspace testkeyspace> |     CREATE
 normaluser |                <keyspace testkeyspace> |      ALTER
 normaluser |                <keyspace testkeyspace> |       DROP
 normaluser |                <keyspace testkeyspace> |     SELECT
 normaluser |                <keyspace testkeyspace> |     MODIFY
 normaluser |                <keyspace testkeyspace> |  AUTHORIZE
```

Figure 4-12. *The new list of permissions for normaluser*

10. Next, let's grant all permissions on all keyspaces to superuser:

```
grant select on all keyspaces to 'superuser';
```

11. And now grant select permission to superuser on the **privileges** table:

```
grant select on anotherkeyspace.privileges to 'superuser';
```

12. Let's view all permissions given on the **privileges** table (see Figure 4-13).

```
list all permissions on privileges;
```

```
 username   | resource                              | permission
------------+---------------------------------------+------------
 normaluser | <table anotherkeyspace.privileges>    |     CREATE
 normaluser | <table anotherkeyspace.privileges>    |      ALTER
 normaluser | <table anotherkeyspace.privileges>    |       DROP
 normaluser | <table anotherkeyspace.privileges>    |     SELECT
 normaluser | <table anotherkeyspace.privileges>    |     MODIFY
 normaluser | <table anotherkeyspace.privileges>    |  AUTHORIZE
  superuser |                   <all keyspaces>     |     SELECT
  superuser | <table anotherkeyspace.privileges>    |     SELECT
```

Figure 4-13. *Permissions on privileges and parent resources*

13. We also can display permissions specific to a resource using the NORECURSIVE specifier (see Figure 4-14).

```
list all permissions on privileges NORECURSIVE;
```

username	resource	permission
normaluser	<table anotherkeyspace.privileges>	CREATE
normaluser	<table anotherkeyspace.privileges>	ALTER
normaluser	<table anotherkeyspace.privileges>	DROP
normaluser	<table anotherkeyspace.privileges>	SELECT
normaluser	<table anotherkeyspace.privileges>	MODIFY
normaluser	<table anotherkeyspace.privileges>	AUTHORIZE
superuser	<table anotherkeyspace.privileges>	SELECT

Figure 4-14. *List of permissions assigned to all users on the privileges table*

14. Let's connect using normaluser and try to view all permissions given on the **privileges** table.

```
list all permissions on anotherkeyspace.privileges;
```

Again, we receive a "not authorized" error (see Figure 4-15). **normaluser** has been given all permissions on the **privileges** table, but it can't view everyone's permissions. However, normaluser can perform other permitted operations mentioned previously.

```
cqlsh> list all permissions on anotherkeyspace.privileges;
Bad Request: You are not authorized to view everyone's permissions
```

Figure 4-15. *No permission given to normaluser on anotherkeyspace*

15. Let's also try having normaluser drop itself and superuser (see Figure 4-16).

```
drop user normaluser;
drop user 'superuser';
```

The result, shown in Figure 4-16, indicates that normaluser can't drop, revoke, or view permissions of other users. But a superuser can!

```
cqlsh> drop user normaluser;
Bad Request: Users aren't allowed to DROP themselves
cqlsh> drop user 'superuser';
Bad Request: Only superusers are allowed to perfrom DROP USER queries
```

Figure 4-16. *Dropping of user is not allowed with non-superuser*

16. So, let's connect with superuser and try to revoke permissions given to a specific user (normaluser, in this case), as follows:

```
revoke all permissions on privileges from normaluser;
list all permissions on privileges;
```

Figure 4-17 shows that the action was successful.

```
cqlsh:anotherkeyspace> list all permissions on privileges;

 username  | resource                            | permission
-----------+-------------------------------------+-----------
 superuser |                   <all keyspaces>   |    SELECT
 superuser | <table anotherkeyspace.privileges>  |    SELECT
```

Figure 4-17. *Permissions of a superuser on the privileges table*

From this series of steps, we conclude:

- A non-superuser can't manage, access, or view other users' permissions.

- Permissions can be given on the keyspace and the specific table/column family.

- A superuser can manage or access permissions of other users.

Accessing system_auth with AllowAllAuthorizer

By default, authentication is disabled in Cassandra and AllowAllAuthenticator is configured as the authenticator in cassandra.yaml configuration. In the previous section we explored various authorization techniques with PasswordAuthenticator. In this section we will explore whether the managing user's permission with AllowAllAuthorizer is permissible or not. This recipe will let you understand what will not work with AllowAllAuthorizer. This is purely for experimentation purposes.

As mentioned above, the system_auth keyspace contains user credentials and permissions details. In this recipe, we will try accessing the system_auth keyspace and managing user credentials without authentication.

1. First let's describe the system_auth keyspace (see Figure 4-18).

```
// describe keyspace.
describe keyspace system_auth;
```

```
cqlsh:system_auth> describe keyspace system_auth;

CREATE KEYSPACE system_auth WITH replication = {
  'class': 'SimpleStrategy',
  'replication_factor': '1'
};

USE system_auth;

CREATE TABLE users (
  name text PRIMARY KEY,
  super boolean
) WITH
  bloom_filter_fp_chance=0.010000 AND
  caching='KEYS_ONLY' AND
  comment='' AND
  dclocal_read_repair_chance=0.000000 AND
  gc_grace_seconds=7776000 AND
  read_repair_chance=0.100000 AND
  replicate_on_write='true' AND
  populate_io_cache_on_flush='false' AND
  compaction={'class': 'SizeTieredCompactionStrategy'} AND
  compression={'sstable_compression': 'SnappyCompressor'};
```

Figure 4-18. *The system_auth keyspace described*

2. Let's explore the **users** table as shown in Figure 4-19.

```
cqlsh:system_auth> select * from users;

 name       | super
------------+-------
 cassandra  | True
```

Figure 4-19. *Default user in the users table*

3. To start, let's try deleting all the data from the **users** table (see Figure 4-20):

    truncate users;

```
cqlsh:system_auth> truncate users;
cqlsh:system_auth> select * from users;
cqlsh:system_auth>
```

Figure 4-20. *Truncated users table*

4. We were able to truncate the **users** table, but now let's see whether we can drop it.

```
drop table users;
```

Figure 4-21 shows that we cannot do this, since it is a system table and restricted by Cassandra for external users.

```
cqlsh:system_auth> drop table users;
Bad Request: Cannot DROP <table system_auth.users>
```

Figure 4-21. *Dropping the users table is not allowed*

5. Next, restart the Cassandra server to verify whether the **users** table is empty (see Figure 4-22):

```
select * from users;
```

```
cqlsh:system_auth> select * from users;

 name      | super
-----------+-------
 cassandra | True
```

Figure 4-22. *Default user cassandra is prepopulated*

Changes made to the system_auth keyspace with AllowAllAuthorizer enabled will be ignored by Cassandra. Hence changes made in step 3 will not be applied and Cassandra will populate default settings as shown in step 2.

6. Let's try to create a user:

```
create user normaluser with password 'normal';
create user normaluser;
```

Creating a user with AllowAllAuthenticator is not permitted in Cassandra as you can see in Figure 4-23. However, as we saw earlier, it is possible with PasswordAuthenticator.

```
cqlsh> create user normaluser with password 'normal';
Bad Request: org.apache.cassandra.auth.AllowAllAuthenticator doesn't support PASSWORD option
cqlsh> create user normaluser;
Bad Request: You have to be logged in and not anonymous to perform this request
```

Figure 4-23. *Trying to create a user with AllowAllAuthenticator while not logged in*

7. We can alter the **system_auth** keyspace for replication strategy as:

```
alter keyspace system_auth with replication = {'class':'SimpleStrategy',
'replication_factor':3};
```

With this recipe, we conclude that we can alter the system_auth keyspace but we cannot create a user with AllowAllAuthenticator (authentication disabled). When authenticating on a Cassandra server using default user credentials, you must configure the consistency level as QUORUM.

Let's discuss SSL encryption and connecting cqlsh and Thrift clients when encryption is enabled.

Preparing Server Certificates

In recent years, social media and Internet-based applications made data accessibility and sharing possible all over the world. SSL protocols are used to send encrypted data over the Internet with secure communication.

For client-server or internode communication over SSL, we need to prepare server certificates. A **keystore** file contains server keys, whereas a **Truststore** contains trusted SSL certificates for all clients or nodes.

Before we start a new recipe, let's discuss some possible errors and resolutions up front. After configuring and upon starting the Cassandra server, if we receive an error such as the following, we need to configure the installed Java version with the Java cryptography extension (JCE).

```
Cannot support TLS_RSA_WITH_AES_256_CBC_SHA with currently installed providers
```

Download the JCE package from:

```
http://www.oracle.com/technetwork/java/javase/downloads/jce-6-download-429243.html (Java 6)
http://www.oracle.com/technetwork/java/javase/downloads/jce-7-download-432124.html (Java 7)
```

Then copy **local_policy.jar** and **US_export_policy.jar** under the $JAVA_HOME/jre/lib/security folder. Since we are done with required configuration, let's explore preparing server certificates with a simple exercise.

1. First, generate a keystore for the server using keytool:

    ```
    keytool -genkey -alias servernode -keystore /home/impadmin/keys/server/server.jks
    -storepass server -keypass server
    ```

Here, /home/impadmin/keys/server is the path to the folder that contains the keystore file. You may change it accordingly. Figure 4-24 shows the basic details to be entered when generating a keystore.

```
What is your first and last name?
  [Unknown]:  localhost
What is the name of your organizational unit?
  [Unknown]:  Authors
What is the name of your organization?
  [Unknown]:  Cassandra
What is the name of your City or Locality?
  [Unknown]:  Noida
What is the name of your State or Province?
  [Unknown]:  UP
What is the two-letter country code for this unit?
  [Unknown]:  IN
Is CN=localhost, OU=Authors, O=Cassandra, L=Noida, ST=UP, C=IN correct?
  [no]:  y
```

Figure 4-24. Inputs provided during generation of the keystore

2. Export the public key part from the keystore file:

    ```
    keytool -export -alias servernode  -file /home/impadmin/keys/publickey.cer -keystore
    /home/impadmin/keys/server.jks -storepass server
    ```

Figure 4-25 shows publickey.cer is stored in the /home/impadmin/keys folder.

```
Certificate stored in file </home/impadmin/keys/publickey.cer>
```

Figure 4-25. *A stored publickey certificate with a full path*

3. Import publickey.cer as a trusted certificate in the server's truststore:

```
keytool -import -v -trustcacerts -alias client -file /home/impadmin/keys/publickey.cer
-keystore /home/impadmin/keys/.truststore -storepass client
```

Figure 4-26 shows the certificate has been successfully added.

```
Serial number: 7464dd0
Valid from: Sat Oct 19 19:13:36 IST 2013 until: Fri Jan 17 19:13:36 IST 2014
Certificate fingerprints:
        MD5:  C5:E0:00:20:96:FE:F2:7C:04:15:25:C9:7C:0C:FF:17
        SHA1: 12:60:B2:37:77:1B:BE:9F:17:13:BB:56:DE:78:5F:CD:F3:E3:CF:5E
        SHA256: 45:E4:28:BB:5F:90:F8:BA:98:0D:5E:66:40:C5:9E:B9:FF:03:DA:1D:7A:ED:AC:9B:17:B0:80:8F:2E:83:B4:42
        Signature algorithm name: SHA1withDSA
        Version: 3

Extensions:

#1: ObjectId: 2.5.29.14 Criticality=false
SubjectKeyIdentifier [
KeyIdentifier [
0000: 77 D8 34 83 1A EF 9C 42   18 65 8B 78 13 5D 33 B7  w.4....B.e.x.]3.
0010: B7 63 ED 9E                                        .c..
]
]

Trust this certificate? [no]:  y
Certificate was added to keystore
[Storing /home/impadmin/keys/.truststore]
```

Figure 4-26. *The client's certificate added successfully to the server's truststore*

With that said, we have prepared server certificates. In next few recipes we will use **truststore** to access the server with trusted certificates. For internode communication, we need to repeat the above recipe on each node and copy each node's public certificate in the **truststore** of each node.

Figure 4-27 shows a graphical representation of client certification preparation and importing with truststore.

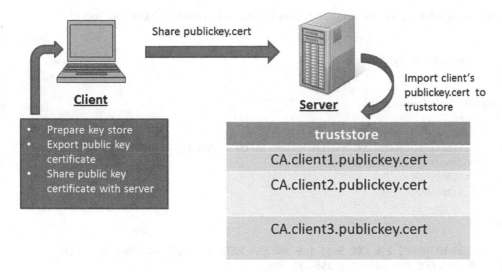

Figure 4-27. *The process to prepare and share a client's public key certificate with the server*

The process of preparing and sharing certificates among Cassandra nodes is the same, and each node has to export and share public key certificates with all other nodes in the Cassandra cluster.

Once the certificate preparation process is complete, you need to enable and connect command line clients with SSL encryption enabled. The next section will discuss this process.

Connecting with SSL Encryption

The previous section discussed connecting with cqlsh having SSL enabled. Let's explore connecting with Cassandra-cli command line client with SSL encryption enabled.

Connecting via Cassandra-cli

Let's enable Cassandra for encrypted client-to-node encryption and connect with the server in the Thrift way (using Cassandra-cli).

1. First we need to configure the server for client-to-node encryption.

    ```
    // enable server
    enabled: true

    // path to keystore
    keystore: /home/impadmin/keys/server.jks

    // keystore password
    keystore_password: server
    ```

The following shows the configuration changes required in `cassandra.yaml` to configure keystore and truststore paths:

```
# enable or disable client/server encryption.
client_encryption_options:
    enabled: false
    keystore: /home/impadmin/keys/server.jks
    keystore_password: server
    # require_client_auth: false
    # Set trustore and truststore_password if require_client_auth is true
    truststore: /home/impadmin/keys/.truststore
    truststore_password: client
    # More advanced defaults below:
    protocol: TLS
    algorithm: SunX509
    store_type: JKS
```

cipher_suites: [TLS ECDHE RSA WITH AES 256 CBC SHA, TLS DHE RSA WITH AES 128 CBC SHA, TLS ECDH ECDSA WITH AES 128 CBC SHA, TLS DHE RSA WITH AES 256 CBC SHA,
SSL DHE RSA WITH 3DES EDE CBC SHA, TLS ECDH RSA WITH AES 256 CBC SHA, SSL RSA WITH RC4 128 SHA,
TLS ECDH ECDSA WITH 3DES EDE CBC SHA, TLS ECDHE RSA WITH RC4 128 SHA,
TLS ECDH ECDSA WITH RC4 128 SHA, TLS ECDHE ECDSA WITH RC4 128 SHA, TLS ECDHE RSA WITH AES 128 CBC SHA, TLS ECDHE ECDSA WITH 3DES EDE CBC SHA,
TLS ECDH RSA WITH RC4 128 SHA, TLS EMPTY RENEGOTIATION INFO SCSV, TLS ECDH RSA WITH 3DES EDE CBC SHA, TLS ECDH RSA WITH AES 128 CBC SHA,
TLS ECDHE ECDSA WITH AES 256 CBC SHA, TLS ECDHE ECDSA WITH AES 128 CBC SHA, TLS DHE DSS WITH AES 256 CBC SHA, TLS RSA WITH AES 128 CBC SHA,
TLS ECDH ECDSA WITH AES 256 CBC SHA, TLS RSA WITH AES 256 CBC SHA, TLS ECDHE RSA WITH 3DES EDE CBC SHA, SSL RSA WITH RC4 128 MD5, TLS DHE DSS WITH AES 128 CBC SHA,
SSI DHF DSS WTTH 3DFS FDF CRC SHA. SSI RSA WTTH 3DFS FDF CRC SHA1

Here **cipher_suites** configuration contains all valid protocols, and it is recommended to keep it as-is while configuring with this application.

You may enable `require_client_auth` to true for client certificate authentication.

2. Start the Cassandra server and connect `cassandra-cli`:

```
$CASSANDRA_HOME/bin/cassandra-cli -h 127.0.0.1 -p 9160 -ts
```

```
/home/impadmin/keys/.truststore -tf org.apache.cassandra.cli.transport.
SSLTransportFactory -tspw client
```

Here, `-ts`, `-tf`, and `-tspw` are truststore, transport factory, and truststore password.

With SSL encryption we can connect to the Thrift client using transport factory and truststore configuration. Next, we will connect cqlsh with SSL encryption enabled.

Connecting via cqlsh

To connect **cqlsh** with encryption enabled, we need to create .cqlshrc file under home directory. Also we can connect cqlsh with require_client_auth=false. Let's connect and configure cqlsh in this recipe.

1. Let's create .cqlshrc file under home directory for SSL-specific configurations.

    ```
    factory = cqlshlib.ssl.ssl_transport_factory

    // path to truststore file
    certfile = /home/impadmin/source/keys/.truststore
    ```

Figure 4-28 shows the configuration changes in the .cqlshrc file.

```
[connection]
hostname = 127.0.0.1
port = 9160
; enable below for ssl
factory = cqlshlib.ssl.ssl_transport_factory


[ssl]
certfile = /home/impadmin/source/keys/.truststore
;; optional - true by default.
validate = true
```

Figure 4-28. *The .cqlshrc configuration changes*

2. That's it! Let's run cqlsh to connect with the Cassandra server:

    ```
    SSL_CERTFILE=/home/impadmin/source/keys/publickey.cer
    SSL_VALIDATE=$CASSANDRA_HOME/bin/cqlsh -3 -t cqlshlib.ssl.ssl_transport_factory
    ```

Connecting via the Cassandra Thrift Client

The following code snippet is the way to connect via a Cassandra Thrift Java-based client with a secure server:

```
TSSLTransportFactory.TSSLTransportParameters params =
        new TSSLTransportFactory.TSSLTransportParameters();
    params.setTrustStore("path to .truststore", "trustore password");

transport = TSSLTransportFactory.getClientSocket("localhost", 7001, 10000, params);
 TProtocol protocol = new TBinaryProtocol(transport);

Cassandra.Client client = new Cassandra.Client(protocol);
```

Here 7011 is the SSL port. The rest is similar to the Cassandra-cli way. Open the client socket using SSL-configured transport parameters (e.g., configure truststore path, password, and SSL transport factory).

That's how we can connect to the Cassandra server via command-line clients and the Java-based Thrift client.

Summary

To summarize, the topics discussed in this chapter are as follows:

- Managing user roles and privileges via super and non-superusers

- Preparing SSL encryption certificates for client-server and internode communication

- Connecting via various clients to secure a Cassandra server.

That ends our discussion around data security and user management with Cassandra.

The next chapter will move closer to large data analytics by discussing **batch processing using Hadoop's MapReduce** algorithm and how to run the MapReduce program over Cassandra using the Cassandra file system as both input and output!

■ ■ ■

MapReduce with Cassandra

So, what's next after discussing data modeling, security, and user role privileges management? With Cassandra query language (CQL), we can definitely manage basic query-based analytics via primary key and secondary indexes and keep data model denormalized as much as possible. But, still, it is possible to perform analytics over a very large chunk of data, in a manner similar to joins, or to persist data into Cassandra after counting specific fields, such as counting the tweets of a particular user account for a given date range. Clearly it's a case of large data analytics, more specifically batch analytics.

In this chapter we will:

- Provide an introduction to MapReduce

- Explore Hadoop

- Discuss HDFS and MapReduce

- Describe integrating Cassandra with MapReduce

Batch Processing and MapReduce

Any form of data, structured or unstructured, would be meaningless unless it gets processed. So far we have discussed various ways to manage and model data volume into Cassandra.

What about running analytics over such archived large data sets? Large data analytics can be divided into two broad categories:

- Batch processing

- Stream processing

In lay terms, batch processing is the execution of one or multiple jobs. These jobs are programmed to require minimal human intervention. Required input/output parameters and resources are preconfigured with jobs. The history of batch processing mechanisms can be traced back to punch cards and mainframe computing.

Let's take an example of a satellite channel application archiving logs for many years. At the end of each year (or maybe half yearly) the provider wants to know how many users of a particular age range have watched specific programs in a primetime slot. Since data volume is huge and totally unstructured, we cannot stream and fit it in-memory to perform such computations. Such data can be largely unrelated to each other and to process these in-batch would require predefined steps for parallel processing.

With respect to large data, batch processing jobs can be categorized in three simple steps:

- extract,

- transform,

- and load

This process is commonly referred to as **ETL**. Various ETL tools such as Ab initio, CloverETL, Pentaho, and Informatica are available for data warehousing and analytics.

Another aspect of ETL systems is analytics. Imagine a system needs to perform big data analytics where data input points are different applications and the system needs to generate a consolidated aggregation report. This is where data would get **extracted** from various input channels and will get computed and transformed before loading it into a database. Figure 5-1 shows an example where ETL based analytics need to extract data from social media channels, financial applications, and server logs. Transformation/computation is done on the engine side and finally the output gets loaded onto the database.

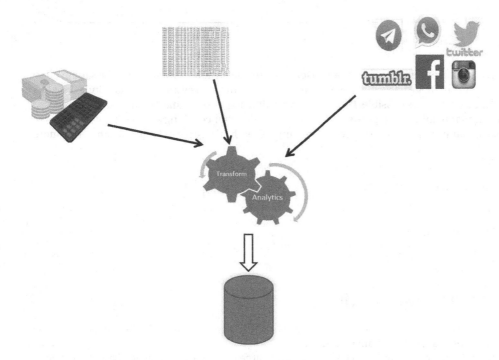

Figure 5-1. *ETL based analytics*

In 2008, Google's research program published a paper "MapReduce: Simplified Data Processing on Large Clusters" that introduced the MapReduce paradigm, though it had been in use since 2003. (You can download this white paper at `http://static.googleusercontent.com/media/research.google.com/en//archive/mapreduce-osdi04.pdf`.) The The MapReduce algorithm is a distributed parallel programming model, comprised of the map and reduce function, and is one solution to business problems that require analytics over heterogeneous forms of massive amounts of data. Since the Google paper was published, various MapReduce implementations have been built. One popular and open-source implementation is **Apache Hadoop**.

Apache Hadoop

Apache Hadoop is a popular open-source library for distributed computation for large-scale data sets. It's a large-scale batch processing infrastructure that is primarily meant to deal with batch analytics over data distributed across hundreds of nodes.

With a MapReduce job, the data set is divided into chunks in such a way that each of these chunks can be processed independently as a map task in a collocated way and then the framework sorts and supplies the output for reduced tasks. The Hadoop framework automatically manages task scheduling and data distribution.

Obviously such a huge amount of data cannot be managed on a single node and it has to be distributed across multiple nodes. Two issues that are clearly visible here are:

- Data Distribution
- Co-located data processing

HDFS

Data distribution with Hadoop is managed with the Hadoop Distributed File System (HDFS). HDFS is one of the submodules for Apache Hadoop project. It is specifically designed to build a scalable system over a less expensive hardware. HDFS is based on master/slave architecture, where a single process known as NameNode runs on the master node and manages all the information about data files and replication across the cluster of nodes.

For more information about HDFS architecture, please refer to `http://hadoop.apache.org/docs/r1.2.1/hdfs_design.html`.

MapReduce

MapReduce implementation is also the answer to another issue: It's a framework for parallel processing of large amounts of data distributed across multiple nodes.

Three primary tasks performed during the MapReduce process are:

- Split and map
- Shuffle and sort
- Reduce and output

Upon submitting a job, input data is split in chunks and assigned as a parallel map task to each mapper. Each mapper generates a distinct key value pair as an output, which is shuffled and sorted by keys and supplied as input to each reducer. Each mapper and reducer running on data node would operate on local data only (data locality). Another added advantage is framework is completely decoupled and custom implementations for mapper, reducer, file format, and record reading.

Let's simplify this with the help of a famous word count problem. Here, the program needs to output a number of occurrences of each word of the input file (see Figure 5-2).

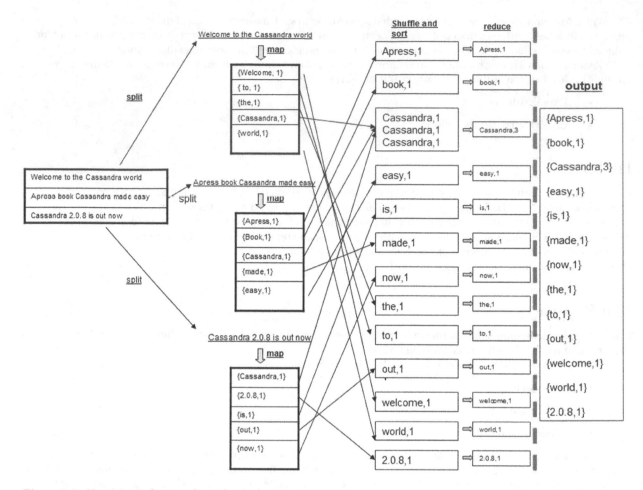

Figure 5-2. *How MapReduce works with word count*

Figure 5-2 is a graphical representation of the MapReduce program of word count for three given lines from a Word file.

1. Welcome to the Cassandra world

2. Apress book Cassandra made easy

3. Cassandra 2.0.8 out now

Here each map task runs locally on data node where the data resides. Then data is implicitly shuffled and sorted by the Hadoop MapReduce framework and finally given to reduce tasks, after the reduction process outputs it into a file.

To understand it on a lower level, you may also refer to the pseudo code representation in Figure 5-3.

```
def map(file)
   for each word in file
      emit(word,1)
```

```
def reduce(word,countArr[])
   var sum
   for( count <- 0 to countArr)
      sum +=count
   emit(sum)
```

Figure 5-3. Word count MapReduce algorithm

Read and Store Tweets into HDFS

Since we have followed Twitter-based examples throughout this book, let's again use real-time tweets for an HDFS/MapReduce example. The reason to start with HDFS is to get started with running MapReduce over default HDFS, and then later explain how it can be executed over an external file system (e.g., Cassandra file system). In this section we will discuss how to set up and fetch tweets about "apress". In our example, we will stream in some tweets about "Apress publications" and store them on HDFS. The purpose of this recipe is to set up and configure Hadoop processes and lay the foundation for the next recipe in the "Cassandra MapReduce Integration" section.

Reading Tweets

To start, let's build a sample Maven-based Java project to demonstrate reading tweets from Twitter. (For more details on using Maven, you may also refer to http://maven.apache.org/plugins/maven-site-plugin/usage.html.)

1. First, let's generate a Maven project:

```
Cmvn archetype:create –DgroupId=com.apress –DartifactId=twitterExample –Dversion=1.0 –Dpackaging=jar
```

2. Let's add a twitter4j dependency (see Figure 5-4) in pom.xml:

```
<!-- Twitter 4j dependency -->
    <dependency>
        <groupId>org.twitter4j</groupId>
        <artifactId>twitter4j-core</artifactId>
        <version>[3.0,)</version>
    </dependency>
```

3. Configure twitter4j.properties (under src/main/resources) for consumer and access credentials:

```
#twitter4j.properties
# Given below keys are masked. Before running DefaultTwitterService, kindly change them according to
your settings.
oauth.consumerSecret=8xxxxxxcxxxxxxxxxxxxxxxxxxxxxxxxxxxxxxx
oauth.accessToken=3ccxxxxxxxxxxxxxxxxxxxxxxxxxxxxxxxxxxxxxxxxxxxxxxx
oauth.accessTokenSecret=hxxxxxxxxxxxxxxxxxxxxxxxxxxxxxxxxxxxxxxxxxxx
oauth.consumerKey=oxxxxxxxxxxxxxxxxxxxxx
```

4.   Let's create a connection using the Twitter API as:

```
twitter = new TwitterFactory().getInstance();
```

■ **Note**   Please refer to `com.apress.chapter5.mapreduce.twittercount.hdfs.ConnectionHandler` for the implementation.

---

5.   Next configure apress as search tokens and fetch tweets:

```
Query query = new Query("apress");

QueryResult result = connection.search(query);
```

6.   Let's store these tweets on a local folder:

```
final String filePath = "tweets"; // you may change it as per your configuration
        File file = new File(filePath);
        FileOutputStream fos = null;
   try
   {
     fos = new FileOutputStream(file);
   }
   catch (FileNotFoundException e)
   {
       // log error
   }
// read and write tweets on local file system
        do
        {
            List<Status> statuses = result.getTweets();

        for (Status status : statuses)
                {
                    StringBuilder sb = new StringBuilder();
                    sb.append(status.getUser().getCreatedAt());
                    sb.append("\001");
                    if (status.getUser() != null)
                    {
                        sb.append(status.getUser().getName());
                        sb.append("\001");
                    }
                    sb.append(status.getText());
                    sb.append("\n");
                    try
                    {
                        fos.write(sb.toString().getBytes());
                    }
```

```
            catch (IOException e)
                {
    // log error
  }
  count++;
     }
        while ((query = result.nextQuery()) != null);

}
```

Running this will store a tweets file in the current directory.

The source code for all the examples used in this chapter is available with the downloads for this book.

The complete source code for the preceding snippet is available at com.apress.chapter5.mapreduce.twittercount. hdfs.DefaultTwitterService.

Storing Tweets into HDFS

Next, store tweets into HDFS. You also need to have Hadoop installed and set up. In this example we will be using Hadoop single node setup, and that's what we'll do first.

A few basic steps to the setup are:

1.   Download the Hadoop tarball distribution:

https://archive.apache.org/dist/hadoop/core/hadoop-1.1.1/hadoop-1.1.1-bin.tar.gz

2.   Extract the tarball into a local folder and run the following command:

$HADOOP_HOME/bin/hadoop namenode -format

Here, $HADOOP_HOME is the directory containing extracted Hadoop binaries.

3.   Modify $HADOOP_HOME/conf/core-site.xml for the Default FS setting:

```
<configuration>
    <property>
        <name>fs.defaultFS</name>
        <value>hdfs://localhost:9000</value>
    </property>
</configuration>
```

4.   Modify $HADOOP_HOME/conf/hdfs-site.xml for the replication setting:

```
<configuration>
    <property>
        <name>dfs.replication</name>
        <value>1</value>
    </property>
</configuration>
```

5. Start the Hadoop process by running the following:

```
$HADOOP_HOME/bin/start-all.sh
```

Now complete the following steps to copy the tweets file:

1. Before we copy the local file on HDFS, let's verify whether all five processes are running. Figure 5-4 shows all those as underlined.

```
17067 TaskTracker
16796 SecondaryNameNode
9633 Jps
16915 JobTracker
16521 NameNode
5243 org.eclipse.equinox.launcher_1.2.0.v20110502.jar
6270 CassandraDaemon
16652 DataNode
```

Figure 5-4. Five Hadoop processes (underlined) running on a local box

2. Copy the tweets file on HDFS:

```
// create /apress/tweetdata on HDFS
$HADOOP_HOME/bin/hadoop fs -mkdir /apress/tweetdata
$HADOOP_HOME/bin/hadoop fs -put $DIR/tweets /apress/tweetdata
```

Here tweets are stored under the HDFS directory: /apress/tweetdata.

3. Once data is copied, you may verify it using the web admin UI (localhost:50070) by clicking on Browse the filesystem, as shown in Figure 5-5. You see data is available using the Hadoop web UI console.

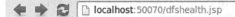

NameNode 'localhost:9000'

Started:	Sun Sep 29 19:33:11 IST 2013
Version:	1.1.1, r1411108
Compiled:	Mon Nov 19 10:48:11 UTC 2012 by hortonfo
Upgrades:	There are no upgrades in progress.

<u>Browse the filesystem</u>
<u>Namenode Logs</u>

Cluster Summary

9 files and directories, 2 blocks = 11 total. Heap Size is 60.19 MB / 888.94 MB (6%)

Configured Capacity	:	220.28 GB
DFS Used	:	1.75 MB
Non DFS Used	:	49.55 GB
DFS Remaining	:	170.73 GB
DFS Used%	:	0 %
DFS Remaining%	:	77.51 %
<u>**Live Nodes**</u>	:	1
<u>**Dead Nodes**</u>	:	0
<u>**Decommissioning Nodes**</u>	:	0
Number of Under-Replicated Blocks	:	0

Figure 5-5. *Hadoop Web UI admin console*

Alternatively, you can run the following command to validate the stored tweet file:

```
$HADOOP_HOME/bin/hadoop fs -lsr /apress/tweetdata
```

We are storing the live Twitter stream on the local file system first just for this sample exercise. With real-time streaming of tweets, you may want to store a live Twitter stream into HDFS using Flume or Scribe. Please refer to https://cwiki.apache.org/FLUME/ or https://github.com/facebook/scribe.

For more information on installation and setup, you can refer to http://hadoop.apache.org/docs/stable/single_node_setup.html.

In this example, we discussed how to store live tweets in HDFS. In the following section, we will explore writing a MapReduce program to store the tweet count of specific users and dates into the Cassandra column family.

Cassandra MapReduce Integration

In this section, we will read tweets (see above section) from HDFS and discuss a MapReduce program to perform the computation of tweets. Finally, reduced output will be stored in the Cassandra **tweetcount** column family. We will discuss MapReduce over Cassandra with two recipes, which are:

- Reading tweets from HDFS and storing tweet counts into Cassandra.

- Reading tweets from Cassandra and storing tweet counts into Cassandra.

For this example, we will be using the Thrift protocol to create a Cassandra schema. As an open-source API for Cassandra, MapReduce integration is available with the Thrift protocol. Sample exercises will demonstrate the MapReduce integration with Cassandra in a CQL3 way.

Reading Tweets from HDFS and Storing Count Results into Cassandra

In this section we will be reading the previously stored tweets file on the HDFS directory /apress/tweetdata (see the preceding section) and storing the tweet count per user and per date in Cassandra. Cassandra provides MapReduce support for both Thrift and CQL3. We will explore both protocols starting with Thrift.

The Thrift Way

Let's explore it with the same Twitter example, where the user and their tweets should be stored locally and sorted by tweet_date.

1. First, we need to prepare the data definition. Let's create a column family **tweetcount** using Cassandra-cli:

```
// create keyspace.
Create keyspace tweet_keyspace;

use tweet_keyspace;

// create column family tweetcount.
create column family tweetcount with comparator='UTF8Type' and key_validation_class = 'UTF8Type' and
column_metadata=[{column_name:'count', validation_class:'Int32Type'}];
```

2. Let's create a MapReduce job. We need to create a Hadoop configuration instance and configure the NameNode host and port:

```
Configuration conf = new Configuration();
conf.set("fs.default.name","hdfs://localhost:9000");    // Change this as per your Hadoop
configuration.
conf.set("mapred.child.java.opts", "-Xms1024m -Xmx2g -XX:+UseSerialGC");
conf.set("mapred.job.map.memory.mb", "4096");
conf.set("mapred.job.reduce.memory.mb", "2048");
conf.set("mapreduce.map.ulimit","1048576");
conf.set("mapred.job.reduce.physical.mb", "2048");
conf.set("mapred.job.map.physical.mb", "2048");
```

■ **Note** You may want to change `fs.default.name` if you're running on a remote machine.

---

3. Let's now configure the MapReduce job for the mapper and reducer:

```
//Mapper configuration
job.setMapperClass(TweetTokenizer.class);
job.setMapOutputKeyClass(Text.class);
job.setMapOutputValueClass(IntWritable.class);
```

```
// Reducer configuration
job.setReducerClass(TweetAggregator.class);
job.setOutputKeyClass(ByteBuffer.class);
job.setOutputValueClass(List.class);
job.setOutputFormatClass(ColumnFamilyOutputFormat.class);
```

4. Next, we need to provide a Cassandra-specific configuration for output mapping:

```
ConfigHelper.setOutputRpcPort(job.getConfiguration(), "9160");
ConfigHelper.setOutputInitialAddress(job.getConfiguration(), "localhost");
ConfigHelper.setOutputPartitioner(job.getConfiguration(), "Murmur3Partitioner");
ConfigHelper.setOutputColumnFamily(job.getConfiguration(), KEYSPACE_NAME, COLUMN_FAMILY);
```

■ **Note**   You can also change the rpc port and initial address in case you are not running on a localhost.

5. Finally, we need to provide input and output paths:

```
FileInputFormat.addInputPath(job, new Path(otherArgs[0]));
job.setOutputFormatClass(ColumnFamilyOutputFormat.class);
```

Here otherArgs[0] is the input path for the HDFS tweet file (e.g., /apress/tweetdata).
Let's have a look at the mapper (TweetTokenizer):

```
public class TweetMapper
{

public static class TweetTokenizer extends Mapper<LongWritable, Text, Text, IntWritable>
{
    private final static IntWritable one = new IntWritable(1);

    /* (non-Javadoc)
     * @see org.apache.hadoop.mapreduce.Mapper#map(KEYIN, VALUEIN, org.apache.hadoop.mapreduce.
       Mapper.Context)
     */
    public void map(LongWritable key, Text value, Context context)
    throws IOException, InterruptedException
    {
        // split into tokens and pass  date and count as key.
        String[] values = StringUtils.split(value.toString(), "\001");

        if(values.length >=2 && values.length <=3)
        {
            context.write(new Text(values[0]), one);  // count on by date.
            context.write(new Text(values[1]), one);  // count on users.
        }

    }
}
}
```

Now, let's have a look at the reducer (TweetAggregator):

```java
public static class TweetAggregator extends  org.apache.hadoop.mapreduce.Reducer<Text,
IntWritable, ByteBuffer, List<Mutation>>
    {
        public void reduce(Text word, Iterable<IntWritable> values, Context context) throws IOException,
            InterruptedException
        {
            int sum = 0;
            for (IntWritable val : values)
                sum += val.get();

            context.write(ByteBufferUtil.bytes(word.toString()), Collections.
            singletonList(getMutation(word, sum)));
        }

        private static Mutation getMutation(Text word, int sum)
        {
            Column c = new Column();
            c.setName(ByteBufferUtil.bytes("count"));
            c.setValue(ByteBufferUtil.bytes(sum));
            c.setTimestamp(System.currentTimeMillis());

            Mutation m = new Mutation();
            m.setColumn_or_supercolumn(new ColumnOrSuperColumn());
            m.column_or_supercolumn.setColumn(c);
            return m;
        }
    }
```

The complete source code of this MapReduce job can be found with the downloads for this book. The executable class is TwitterHDFSJob (com.apress.chapter5.mapreduce.twittercount.hdfs package). You also can refer to the README.txt (under src/main/resources) file for further instructions about setting the database and running this MapReduce job.

After successfully executing the job, the output in the tweetcount column family will be as shown here:

```
[default@tweet_keyspace] list tweetcount;
Using default limit of 100
Using default cell limit of 100
-------------------
RowKey: Mon May 11 06:16:04 IST 2009
=> (name=count, value=334, timestamp=1407569960904)
-------------------
RowKey: Hazem Saleh
=> (name=count, value=334, timestamp=1407569960798)
-------------------
RowKey: cessprin
=> (name=count, value=334, timestamp=1407569960990)
-------------------
```

```
RowKey: Thu Feb 23 00:24:16 IST 2012
=> (name=count, value=334, timestamp=1407569960966)
-------------------
RowKey: Hunter Scott
=> (name=count, value=334, timestamp=1407569960803)

22 Rows Returned.
```

The sample output of the stored tweetcount includes a total of 22 rows. The row key is either tweet date or user name and contains a column with the name as count and its value.

The CQL3 Way

In previous chapters we discussed the differences between and interoperability issues with CQL3 and Thrift. Cassandra provides support for CQL-compatible input and output format classes for MapReduce. Please note that these implementations are still based on Thrift but not the native CQL3 driver. In this example, we will be using CQLOutputFormat for writing output in the CQL3 column family. We know that column families created via CQL3 are not visible with Thrift, so let's explore how we can run the MapReduce over CQL3 table/column families. Running the preceding MapReduce program with CQL3 requires very few changes. We need to define the table in CQL3 format and change the Hadoop job configuration to point to the CQL3-based output format. Let's discuss these changes as follows:

1. First, you need to create the table tweetcount:

```
create table tweetcount_cql (key text primary key, count int);
```

2. Changes required at the Hadoop job level are:

```
// set update CQL and row key
        job.getConfiguration().set("row_key", "key");
        String query = "UPDATE " + KEYSPACE_NAME + "." + COLUMN_FAMILY + " SET count = ? ";
        CqlConfigHelper.setOutputCql(job.getConfiguration(), query);
//set cql outputformat class
        job.setOutputFormatClass(CqlOutputFormat.class);
```

3. The CQL3-based aggregator is:

```
public static class TweetCQLAggregator extends org.apache.hadoop.mapreduce.Reducer<Text,
IntWritable, Map<String,ByteBuffer>, List<ByteBuffer>>
    {

        private static Map<String,ByteBuffer> keys = new HashMap<>();

        /* (non-Javadoc)
* @see org.apache.hadoop.mapreduce.Reducer#reduce(KEYIN, java.lang.Iterable,org.apache.hadoop.
mapreduce.Reducer.Context)
        */
        public void reduce(Text word, Iterable<IntWritable> values, Context context) throws
        IOException,
                InterruptedException
```

```
        {
            int sum = 0;
            for (IntWritable val : values)
                sum += val.get();

            System.out.println("writing");
            keys.put("key", ByteBufferUtil.bytes(word.toString()));
            context.write(keys, getBindVariables(word, sum));
        }

        private List<ByteBuffer> getBindVariables(Text word, int sum)
        {
            List<ByteBuffer> variables - new ArrayList<ByteBuffer>();
            variables.add(Int32Type.instance.decompose(sum));
            return variables;
        }
    }

}
```

Finally, after running the HDFS job (see TwitterHDFSCQLJob, source referenced after the output), we can see the output using the cqlsh client:

```
select * from tweetcount_cql;
```

key	count
Mon May 11 06:16:04 IST 2009	334
Hazem Saleh	334
cessprin	334
Thu Feb 23 00:24:16 IST 2012	334
Hunter Scott	334
Tue Sep 29 22:01:16 IST 2009	334
Elena Vielva Gómez	334
Sun Dec 18 20:13:39 IST 2011	334
Alejandro || Serras	334
Tue Jul 06 18:01:47 IST 2010	334
Wed Mar 07 05:11:11 IST 2012	334
Wed Nov 07 14:04:18 IST 2012	334
Oseias Moraes	334
Adeesh Fulay	334
Sat Nov 05 21:29:20 IST 2011	334
Sun Jul 24 21:44:17 IST 2011	334
Manthita.	334
ebooksdealofdaybot	334
Wed Apr 23 19:19:49 IST 2014	334
The News Selector	6680
Louise Corrigan	334
Mon Mar 03 01:19:17 IST 2014	6680

22 Rows Returned.

The complete source code is available with the downloads for this book, and classes discussed are

- `com.apress.chapter5.mapreduce.twittercount.hdfs.TwitterHDFSCQLJob`
- `com.apress.chapter5.mapreduce.twittercount.hdfs.TweetAggregator`

In next section we will discuss using Cassandra as an input and output format for MapReduce.

Cassandra In and Cassandra Out

Let's discuss running a MapReduce where input will be fetched from Cassandra and output will also get stored in Cassandra.

So far we have seen that the MapReduce job execution is possible over default HDFS and over an external file system such as Cassandra. You must be wondering which file system to adopt and why? Well it depends on the use case. For example, if an application has already been built using various Cassandra features, it's better to keep its MapReduce base batch analytics to be implemented in Cassandra. There can be use cases where HDFS has already been used for storing raw data and the user might not agree with migration but still want to run a few MapReduce jobs and store output into Cassandra. Similarly the user may want to migrate away from HDFS and its ecosystem (Hive, Pig, and so forth) to a single database solution (i.e., Cassandra). One big difference we must remember is that HDFS is a distributed file system, whereas Cassandra is a distributed database. Cassandra is fault-tolerant and doesn't have a single point of failure whereas HDFS is not. Another key difference is that Hadoop is master-slave architecture whereas Cassandra is peer to peer. Since the solutions build over Cassandra are scalable and use Cassandra's specific features (such as secondary indexes, composite columns, etc.), we still might need to perform batch analytics using MapReduce over Cassandra. In this recipe, we discuss the same tweet count example using Cassandra as both the input and output format.

The program takes a user name as an input argument (the default user value is **mevivs**), for which a number of tweets is calculated.

1. We need to prepare the data definition first. Let's create a keyspace `tweet_keyspace` and column families **tweetstore** and **tweetcount.** Here `tweetstore` will store raw tweets, whereas the count for a specific user will be stored in the `tweetcount` column family via cqlsh.

```
// create keyspace.
create keyspace tweet_keyspace with replication={'class': 'SimpleStrategy', 'replication_factor:3};
use tweet_keyspace;

// create input column family.
create table tweetstore(tweet_id timeuuid PRIMARY KEY, user text, tweeted_at timestamp, body text);

// update column family from Cassandra-cli(thrift way) to enable index over user.
create column family tweetstore with column_metadata=[{column_name:'user', validation_
class:'UTF8Type', index_type: KEYS},{column_name:'body', validation_class: 'UTF8Type'},
{column_name:'tweeted_at',validation_class: 'DateType'}]

// create output column family via Cassandra-cli.
create column family tweetcount with comparator='UTF8Type' and key_validation_class = 'UTF8Type'
and column_metadata=[{column_name:'count', validation_class:'Int32Type'}];
```

2. Let's populate some rows in tweetstore using **cqlsh** client:

```
use tweet_keyspace;

    // copy command
    copy tweetstore from STDIN;
```

The following is the console output of copying from STDIN:

```
cqlsh:tweet_keyspace> copy tweetstore from STDIN;
[Use \. on a line by itself to end input]

[copy]  now(),bama meeting with European allies on Ukraine http://t.co/1Ik3qBOPxC from #APress #tns,
2013-10-20,mevivs

[copy] now(),nergdahl uproar halts plan for return celebration http://t.co/bywaBAoWTP from #APress
#tns,2012-09-22,Chrisk

[copy] now(),booksdealofdaybot^A[Apress]?Android Apps Security?http://t.co/5XLSObkk9f,2014-03-24,mevivs
[copy] \.

3 rows imported in 6 minutes and 35.900 seconds.
```

3. Let's create a MapReduce job with this mapper and reducer configuration:

```
Configuration conf = new Configuration();
String[] otherArgs = new GenericOptionsParser(conf, args).getRemainingArgs();
Job job = new Job(conf, "tweet count");

job.setJarByClass(TwitterCassandraJob.class);

// mapper configuration.
job.setMapperClass(TweetMapper.class);
job.setMapOutputKeyClass(Text.class);
job.setMapOutputValueClass(IntWritable.class);
job.setInputFormatClass(ColumnFamilyInputFormat.class);

// Reducer configuration
job.setReducerClass(TweetAggregator.class);
job.setOutputKeyClass(ByteBuffer.class);
job.setOutputValueClass(List.class);
job.setOutputFormatClass(ColumnFamilyOutputFormat.class);
```

4. Next, we need to configure the MapReduce job for input family and format configuration:

```
// Cassandra input column family configuration
ConfigHelper.setInputRpcPort(job.getConfiguration(), "9160");
ConfigHelper.setInputInitialAddress(job.getConfiguration(), "localhost");
ConfigHelper.setInputPartitioner(job.getConfiguration(), "Murmur3Partitioner");
ConfigHelper.setInputColumnFamily(job.getConfiguration(), KEYSPACE_NAME, INPUT_COLUMN_FAMILY);

job.setInputFormatClass(ColumnFamilyInputFormat.class);
```

5. Since we need to fetch records for a specific indexed column (**user**), we will be using the Thrift Slice API to configure the MapReduce job with the mapper class for filtered streaming.

```
// Create a slice predicate

SlicePredicate slicePredicate = new SlicePredicate();
slicePredicate.setSlice_range(new SliceRange(ByteBufferUtil.EMPTY_BYTE_BUFFER, ByteBufferUtil.EMPTY_
BYTE_BUFFER, false, Integer.MAX_VALUE));

// Prepare index expression.
IndexExpression ixpr = new IndexExpression();ixpr.setColumn_name(ByteBufferUtil.bytes(COLUMN_NAME));
ixpr.setOp(IndexOperator.EQ);
ixpr.setValue(ByteBufferUtil.bytes(otherArgs.length > 0 && !StringUtils.isBlank(otherArgs[0])?
otherArgs[0]: "mevivs"));

List<IndexExpression> ixpressions = new ArrayList<IndexExpression>();
ixpressions.add(ixpr);

ConfigHelper.setInputRange(job.getConfiguration(), ixpressions);
ConfigHelper.setInputSlicePredicate(job.getConfiguration(), slicePredicate);
```

6. Next, we need to configure the MapReduce job for the output family and format configuration:

```
// Cassandra output family configuration.
ConfigHelper.setOutputRpcPort(job.getConfiguration(), "9160");
ConfigHelper.setOutputInitialAddress(job.getConfiguration(), "localhost");
ConfigHelper.setOutputPartitioner(job.getConfiguration(), "Murmur3Partitioner");
ConfigHelper.setOutputColumnFamily(job.getConfiguration(), KEYSPACE_NAME, OUTPUT_COLUMN_FAMILY);

job.setOutputFormatClass(ColumnFamilyOutputFormat.class);
```

7. Let's have a look at the mapper (TweetMapper):

```
public class TweetMapper extends Mapper<ByteBuffer, SortedMap<ByteBuffer, Column>, Text, IntWritable>
{
    static final String COLUMN_NAME = TwitterCassandraJob.COLUMN_NAME;

    private final static IntWritable one = new IntWritable(1);

    /* (non-Javadoc)
     * @see org.apache.hadoop.mapreduce.Mapper#map(KEYIN, VALUEIN, org.apache.hadoop.mapreduce.
       Mapper.Context)
     */
    public void map(ByteBuffer key, SortedMap<ByteBuffer, Column> columns, Context context) throws
    IOException,
            InterruptedException
    {
```

```
        Column column = columns.get(ByteBufferUtil.bytes(COLUMN_NAME));
        String value = ByteBufferUtil.string(column.value());
        context.write(new Text(value), one);
    }
}
```

It simply reads a specific column (e.g., user from streaming columns) and writes a count for each column value. We will be using the same reducer (TweetAggregator) to perform reduce operations.

The complete source code of this MapReduce job can be found with the downloads for this book. The executable class is TwitterCassandraJob (com.apress.chapter5.mapreduce.twittercount.cassandra package). You may also refer to README.txt and db.txt (under src/main/resources) file for further instructions.

After successfully executing the job, the output in the tweetcount column family is shown in Figure 5-6.

```
[default@tweet_keyspace] list tweetcount;
Using default limit of 100
Using default column limit of 100
-------------------
RowKey: mevivs
=> (column=count, value=2, timestamp=1380554854391)

1 Row Returned.
Elapsed time: 139 msec(s).
```

Figure 5-6. *Counts specific to user mevivs are stored in tweetcount*

Complete source code for this recipe is available under com.apress.chapter5.mapreduce.twittercount.cassandra folder in the downloads for this book.

Stream or Real-Time Analytics

Batch processing frameworks is a good fit for a write-once/read-everywhere paradigm. But for continuous updates to the data set, any in-process Hadoop job will not pick those data updates and would require a rerun.

Real-time analytics would require processing and analyzing a massive amount of data as it enters the system. Applications such as stock market trading and dynamic predictive analysis would require providing analytics in real time as the data gets processed on to the system.

In the last year or so, there has been significant interest in building such a real-time analytics application. As a result, there are number of new frameworks, such as storm, Samaza, and Kafka. We will discuss their integration in subsequent chapters.

Summary

As mentioned previously, all the code snippets shared in this chapter are available as complete source code with the downloads for this book. The download also contains db.txt and README.txt for instructions about configuring and running these Java programs and the Cassandra data definition used in these examples.

In this chapter, we discussed HDFS and MapReduce and integrating them with Cassandra. Running MapReduce is one solution for large data analytics. However, to meet more complex analytics requirements, it might be necessary to use built-in APIs that can build and execute such MapReduce programs automatically. Two of the most popular tools are Hive and Pig. The next chapter will discuss using Hive and Pig and their integration with Cassandra.

CHAPTER 6

Data Migration and Analytics

In the previous chapter, we discussed the benefits of and requirements for running batch analytics over Cassandra via Hadoop MapReduce. We can easily utilize Hadoop MapReduce's pluggable architecture to implement MapReduce even with custom implementations. Let's talk about a few of the published use cases as follows:

- NetApp collects and analyzes system diagnostic-related data for improving the quality of client site deployed systems.

- The leading health insurance provider collects and processes millions of claims per day. Data ingestion is approximately 1 TB per day.

- Nokia collects and analyzes data related to various mobile phones. The expected data volume to deal with is about 600 TB with both structured and unstructured forms of data.

- Etsy, an online marketplace for handmade items, needs to analyze billions of logs for behavior targeting and building search-based recommendations.

Storage of large amounts of data is a significant issue, and with Cassandra we can achieve faster ingestion. What about analyzing these large datasets? CQL3 comes in very handy as an SQL-like interface, but to process and analyze in parallel batches, we need to implement MapReduce-like algorithms. Previous chapters cover implementing MapReduce basics and implementing in Java. But in a few cases we prefer ready-to-use and easy-to-integrate solutions!

In this chapter we will discuss

- Apache Pig setup and basics

- Integrating Pig with Cassandra

- Importing data into Cassandra

- Apache Hive setup and basics

- Hive external and internal tables

- Hive with Cassandra

- Introducing Sqoop

- Sqoop with Cassandra

Data Migration and Analytics

Data migration is the process of data transfer among multiple storage systems. With big data problems, the key point for data migration can be upgrading from one storage system (e.g., an RDBMS) to another one (e.g., Cassandra) for scalability and performance. With the recent emergence of many NoSQL databases, data migration has been an important process to evaluate and implement solutions with the preferred database.

As discussed in the above use cases, data analytics of structured or unstructured forms of data is another issue. Exploring new business domains and dealing with data that has been considered of no use due to known limitations are a few important factors of large data analytics.

Data science is an emerging field these days involving the study of the filtering and extraction of information from raw data. It's an integral part of artificial and competitive intelligence, and data mining and analysis are other practices related to data science.

The process of data mining is to extract and define a pattern after studying big data sets using various artificial intelligence and natural language process-related techniques. Risk analysis, behavior targeting, and forecasting are a few of the big data use cases that require data mining and pattern analysis.

The debate between front-end versus back-end implementation has been underway for many years. Programmers and database architects have their own views on this topic. People with experience in Java or other technologies can easily write such MapReduce implementations. Similarly developers who are well-versed with databases always prefer to implement such algorithms on the database side. Let's try to visualize how a MapReduce job would differ for back-end and application programmers.

Figure 6-1 shows a MapReduce job execution of counting tweets by date on the business layer (programmatically) and storing the output on a database. Here the MapReduce algorithm implementation is on the application level. Such implementations can be Java or any other language.

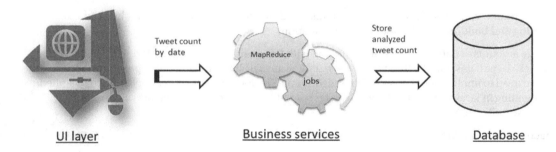

Figure 6-1. *Image depicting standard implementation of tweet analytics at business layer*

Another implementation for the same use case can be that backend programmers or data scientists prefer to write MapReduce scripts or functions on the database side, as shown in Figure 6-2.

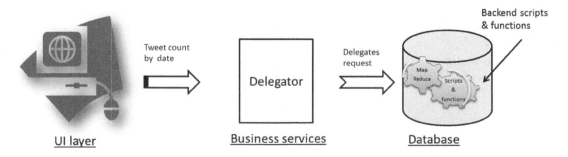

Figure 6-2. *An image depicting implementation of tweet analytics at the database layer*

Developing such MapReduce programs for almost everything is definitely not maintainable and is time-consuming, as well. That's where an SQL-like language or writing a simple script can be really helpful. Database administrators or scientists might prefer a database-based approach (Figure 6-2) for quick implementation. With this we can conclude that there is a space for requirement of such tools. For example, these tools come in very handy to implement a scheduled job which requires extraction over large data sets and stores output.

This comes with a question, are there any tools or open source libraries that are available as a ready-to-use solution. Tools like Apache Hive, Apache Pig, and Sqoop come in very handy for such needs. Before we explore each of these tools, let's go over a brief introduction to each of these.

Apache Pig allows end programmers to write MapReduce implementations in the form of scripts. Apache Pig simply translates this Pig script into Hadoop-compatible MapReduce implementations. There are functions and data type support available with Apache Pig that provide easy and reusable integration to quickly write Pig-powered MapReduce implementations. People building data pipeline or ETL-type solutions prefer to use Pig, as it is procedural but not declarative. Since it is not declarative, you can create checkpoints and plug in custom code at any point of the workflow.

Apache Hive enables users to manage and analyze large data sets using SQL-like query language. SQL has been popular and widely used across the industry. It enables programmers to quickly adopt Hadoop and HBase big data platforms by providing a query-like interface, namely Hive Query Language (Hive QL). Generally it is used for ad-hoc SQL-based analytics. With Hive QL we can perform various DDL and DML operations in an SQL manner. Data definition language (DDL) is used for performing tasks like creating and altering tables, and data manipulation language (DML))) is used to do things like inserting and deleting records. DDL and DML semantics are similar to SQL's. You can refer to `https://cwiki.apache.org/confluence/display/Hive/GettingStarted#GettingStarted-DDLOperations` for more information about DDL. Hive's data partitioning and external table support gives users an added advantage to declare and analyze data over external file systems using Hive. We will cover this in a later part of this chapter.

Sqoop means SQL to Hadoop. Solutions built over RDBMS are not scalable and the user is looking forward to migrate on big data powered solutions. The first priority is migrating existing production data to another database or file system. This is where Apache Sqoop comes in very handy and can help to easily migrate data from one database to another.

Now, let's explore each one of these tools in detail. We'll start with Apache Pig.

Apache Pig

Apache Pig is a platform that provides a simple scripting language known as Pig Latin to build the MapReduce program in an abstract way. Initially it was developed as part of Yahoo's research-related work but later moved to Apache incubation in 2007. It is named as Pig as it can ingest/read in almost any format.

Setup and Installation

There are two possible ways to set up the Apache Pig distribution. One can either download the tarball manually or set up a third-party binary distribution (`.deb` or `.rpm`) over Linux boxes.

For Windows, it requires downloading and setting up a Linux-like environment using cygwin. Cygwin is Unix-like environment and command-line interface for Windows. Cygwin provides native integration of Windows-based applications, data, and other system resources with applications, software tools, and data of the Unix-like environment. For more details, you can refer to `http://en.wikipedia.org/wiki/Cygwin.You` can download and set up cygwin from `www.cygwin.com/install.html`.

In this chapter, we will set up using a tarball distribution for setup and other configuration. To set up this over your local box, let's follow the following sequence:

1. First create a folder:

    ```
    mkdir pig-dist
    cd pig-dist
    ```

2. Then, download the tarball distribution:

    wget http://mirrors.ibiblio.org/apache/pig/pig-0.12.1/pig-0.12.1.tar.gz

---

■ **Note**   During the writing of this chapter, the latest tarball installation is 0.12.1. You may change it for the latest version accordingly.

---

3. Extract the downloaded tarball using the following command:

    ```
    tar -xvf pig 0.12.1.tar.gz
    ```

After step 3, the `pig-dist` folder will have the extracted distribution under the `pig-0.12.1` folder as shown in Figure 6-3.

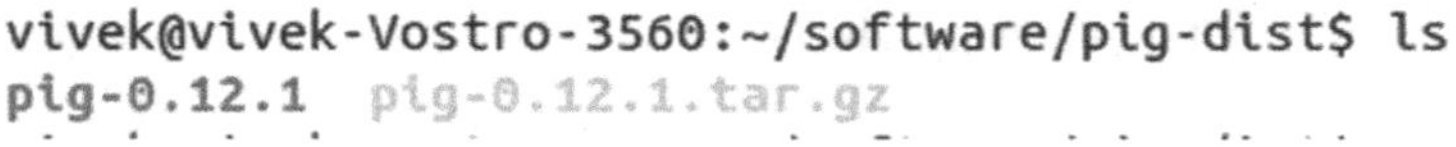

```
vivek@vivek-Vostro-3560:~/software/pig-dist$ ls
pig-0.12.1  pig-0.12.1.tar.gz
```

Figure 6-3. *The extracted tarball distribution in a local folder*

Understanding Pig

Figure 6-4 shows Pig architecture and various components of Apache Pig. It depicts how Pig scripts get compiled and converted into a MapReduce job. With Pig's architecture, a written Pig script would get compiled into MapReduce programs and submitted by the Hadoop job manager as a MapReduce job. The Grunt command shell is a command-line interface for running Pig scripts and the Pig API can be used for building custom implementations.

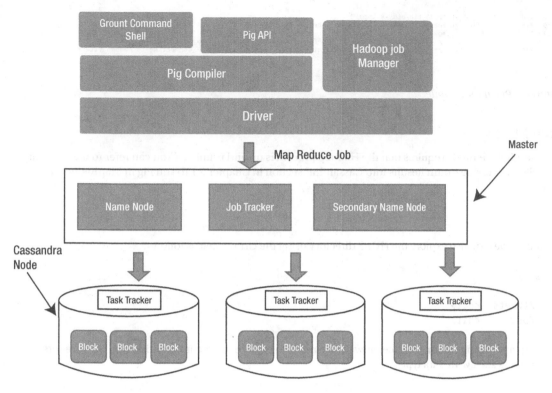

Figure 6-4. *Apache Pig components and MapReduce job transformation*

Before we start on examples, let's discuss these two modes and then a few important Pig commands and data types that we will be using in upcoming exercises in this chapter.

Pig Execution Modes

With Apache Pig, two execution modes are available:

- Local
- MapReduce

Local Mode

Pig has an interactive shell, and we can start Pig's execution in local mode, which will require running the following command (see Figure 6-5):

```
pig -x local
```

```
vivek@vivek-Vostro-3560:~/software/pig-0.12.1$ bin/pig -x local
2014-06-07 16:11:09,398 [main] INFO  org.apache.pig.Main - Apache Pig version 0.12.1 (r1585011) compiled Apr 05 2014, 01:41:34
2014-06-07 16:11:09,399 [main] INFO  org.apache.pig.Main - Logging error messages to: /home/vivek/software/pig-0.12.1/pig_1402137669391.log
2014-06-07 16:11:09,432 [main] INFO  org.apache.pig.impl.util.Utils - Default bootup file /home/vivek/.pigbootup not found
2014-06-07 16:11:09,731 [main] INFO  org.apache.pig.backend.hadoop.executionengine.HExecutionEngine - Connecting to hadoop file system at: file
:///
grunt> ▮
```

Figure 6-5. *Starting Pig in local mode*

MapReduce Mode

Running Pig in MapReduce mode requires that the Hadoop process is up and running. (You can refer to the "Reading Tweets from HDFS and Storing Count Results into Cassandra" section in Chapter 5.) To run Pig in MapReduce mode, we need to run

```
$ pig
```

Once connected we can also explore the HDFS directory using the Grunt shell as follows:

```
grunt> cd hdfs:///
grunt> ls
hdfs://0.0.0.0/tweets   <dir>
hdfs://0.0.0.0/usr      <dir>
```

With this setup and configuration in place, now we are ready to explore other configurations and feature sets provided by Pig. So let's start with data types.

Data Types

Data types supported by Pig can be categorized as *simple* data types and *complex* data types.

Simple Data Types

Simple data types, such as String, int, and so on, are those generally available with most databases and are most often used. Table 6-1 shows all the simple data types supported by Pig.

Table 6-1. *Pig-Supported Simple Data Types*

Data type	Description
charArray	UTF-8 encoded string (e.g., Cassandra)
ByteArray	Byte array (blob)
Double	64-bit precision format (e.g., 11.9)
Float	32-bit precision format (e.g., 10.2f)
Int	32-bit signed integer (e.g., 100)
Long	64-bit signed integer (e.g., 50L)

Complex Data Types

Complex data types, such as map which holds the key-value pair, prefer some kind of predefined data structure format. Table 6-2 shows all three complex data types supported by Pig.

Table 6-2. *Pi- Supported Complex Data Types*

Data Type	Description
Tuples	A tuple is an ordered set of data. A tuple can be thought of as a row with multiple fields and these fields can be of any type and may or may not have data. Tuples are enclosed with ().
	For example, (Vivek,Apress,6) depicts a tuple having author name, publishing house, and chapter number fields.
Bags	A bag is a collection of tuples. A bag can have duplicate tuples and each tuple may differ in the number of fields and their data types.
	For example, this is a bag with multiple tuples:
	{(Vivek,Apress,6),(Chris,Apress,Reviewer,6),{Melissa,Apress,Coordinator,6), {Brian,Apress,Tech reviewer)}
Map	A map is a collection of key value pairs. Each key value pair is delimited by #.
	For example, [name#vivek,email#vivek.mishra@nomail.com] is a map containing two key value pairs. The first element has name as key and vivek as value, whereas the second one has email as key and vivek.mishra@nomail.com as value.

Pig Functions

Pig comes with several built-in functions. Users can also implement custom user-defined functions (UDFs). Built-in pig functions can further be categorized as

- Eval functions
- Math functions
- String functions
- Store functions
- Bag/tuple functions

In this section, let's look at some of the most commonly used Pig functions.

PigStorage

The default function to load data in UTP-8 format. Examples of PigStorage are in the next two sections.

LOAD

The LOAD function loads data from the file system. By default it is used in conjunction with the PigStorage function as shown in the following steps:

1. Create a sample tweets.txt on the local file system in a folder (in this example it is /home/vivek). The file will contain data in the following format:

   ```
   apress_team|technology|A whole bunch of new technology books about to come.
   Watch this space!
   ```

2. Run the following command to load the tweets file using PigStorage with the Grunt shell:

   ```
   pipe_input = LOAD '/home/vivek/input/tweets.txt' USING PigStorage('|');
   ```

3. To dump the output of pipe_input (as shown in Figure 6-6), run the following command:

   ```
   dump pipe_input;
   ```

```
2014-06-14 23:17:08,499 [main] INFO  org.apache.pig.backend.hadoop.executionengine.mapReduceLayer.MapReduceLauncher - Success!
2014-06-14 23:17:08,501 [main] WARN  org.apache.pig.data.SchemaTupleBackend - SchemaTupleBackend has already been initialized
2014-06-14 23:17:08,503 [main] INFO  org.apache.pig.backend.hadoop.mapreduce.lib.input.FileInputFormat - Total input paths to process : 1
2014-06-14 23:17:08,503 [main] INFO  org.apache.pig.backend.hadoop.executionengine.util.MapRedUtil - Total input paths to process : 1
(apress_team,technology,A whole bunch of new technology books about to come. Watch this space!)
grunt> ▮
```

Figure 6-6. *Output of the MapReduce program*

STORE

This function is used to load intermediate or computed Pig script results on output results in the file system. In the following example, we are reading tweets.txt containing fields delimited by '|' and storing it as a file containing fields delimited by ',' (see Figure 6-7).

```
pipe_input = LOAD '/home/vivek/input/tweets.txt' USING PigStorage('|') as (screen_name:chararray,cat
egory:chararray,body:chararray);
csv_output = Store input into '/home/vivek/output/tweets.csv' USING PigStorage(',');
```

```
Success!

Job Stats (time in seconds):
JobId    Alias    Feature Outputs
job_local_0003  pipe_input    MAP_ONLY      /home/vivek/output/tweets.csv,

Input(s):
Successfully read records from: "/home/vivek/input/tweets.txt"

Output(s):
Successfully stored records in: "/home/vivek/output/tweets.csv"

Job DAG:
job_local_0003

2014-06-14 23:21:00,261 [main] INFO  org.apache.pig.backend.hadoop.executionengine.mapReduceLayer.MapReduceLauncher - Success!
```

Figure 6-7. *The output of reading and storing tweets in a csv file*

Output of the preceding command is shown in Figure 6-7.

Upon running the preceding command, a new file (part-m-00000) will be written in the /home/vivek/output folder which will contain comma-separated values as

```
apress_team,technology,A whole bunch of new technology books about to come. Watch this space!
```

Such files are output files generated by the MapReduce job executed with the Pig script. For more details about MapReduce, please refer to the previous chapter.

Please note that here we have used the default storage function PigStorage(), but readers may create their own UDFs and can store/load by using them. For example, in the case of Cassandra, to load data in the Cassandra file system, the CSVStorage and CassandraStorage functions will be used. We will discuss Cassandra's Pig-specific functions in coming exercises.

FILTER

FILTER is used for rows/tuple selection based on the provided condition. Let's describe pipe_input and filter it by screen_name for the value apress_team:

```
describe pipe_input;
filter_by_name = FILTER pipe_input by screen_name matches 'apress_team';
```

Running this command will filter pipe_input for screen_name instances with the value apress_team (see Figure 6-8).

```
2014-06-14 23:34:30,024 [main] INFO  org.apache.pig.backend.hadoop.executionengine.mapReduceLayer.MapReduceLauncher - Success!
2014-06-14 23:34:30,025 [main] WARN  org.apache.pig.data.SchemaTupleBackend - SchemaTupleBackend has already been initialized
2014-06-14 23:34:30,028 [main] INFO  org.apache.hadoop.mapreduce.lib.input.FileInputFormat - Total input paths to process : 1
2014-06-14 23:34:30,028 [main] INFO  org.apache.pig.backend.hadoop.executionengine.util.MapRedUtil - Total input paths to process : 1
(apress_team,technology,A whole bunch of new technology books about to come. Watch this space!)
```

Figure 6-8. *Output of the FILTER command*

FOREACH

Use this function to iterate over result bags and transform into output or intermediate results (see Figure 6-9).

```
similar_result = FOREACH pipe_input GENERATE(*);
result_with_screen_name_only=FOREACH pipe_input GENERATE screen_name;
dump result_with_screen_name_only;
```

```
2014-06-14 23:37:58,504 [main] INFO  org.apache.pig.backend.hadoop.executionengine.mapReduceLayer.MapReduceLauncher - Success!
2014-06-14 23:37:58,505 [main] WARN  org.apache.pig.data.SchemaTupleBackend - SchemaTupleBackend has already been initialized
2014-06-14 23:37:58,507 [main] INFO  org.apache.hadoop.mapreduce.lib.input.FileInputFormat - Total input paths to process : 1
2014-06-14 23:37:58,507 [main] INFO  org.apache.pig.backend.hadoop.executionengine.util.MapRedUtil - Total input paths to process : 1
(apress_team)
```

Figure 6-9. *Running FOREACH to generate intermediate results*

TOTUPLE

TOTUPLE is a function to generate tuples. For example we can use TOTUPLE to generate a tuple of column name and value as shown in Figure 6-10.

```
screen_tuple = FOREACH pipe_input GENERATE TOTUPLE('screen_name',screen_name);
```

```
2014-06-14 23:43:11,087 [main] INFO  org.apache.pig.backend.hadoop.executionengine.mapReduceLayer.MapReduceLauncher - Success!
2014-06-14 23:43:11,088 [main] WARN  org.apache.pig.data.SchemaTupleBackend - SchemaTupleBackend has already been initialized
2014-06-14 23:43:11,089 [main] INFO  org.apache.hadoop.mapreduce.lib.input.FileInputFormat - Total input paths to process : 1
2014-06-14 23:43:11,089 [main] INFO  org.apache.pig.backend.hadoop.executionengine.util.MapRedUtil - Total input paths to process : 1
((screen_name,apress_team))
```

Figure 6-10. *Depicts generated tuple having screen_name as apress_team*

Previously, we discussed a few basic Pig Latin commands. Let's explore Apache Pig with more sample exercises. For all the exercises in this chapter we will be referring to the **tweets** file which contains tweets about **apress.** You can download the sample tweets file from the attachment folder (`datafiles`). This file contains tweet data, screen names, and the tweet body delimited by '\001' (see Figure 6-11).

```
Mon Mar 03 01:19:17 IST 2014░The News Selector░Obama meeting with European allies on Ukraine http://t.co/6xU6JfMsXI from #APress #tns
Mon Mar 03 01:19:17 IST 2014░The News Selector░Bergdahl uproar halts plan for return celebration http://t.co/BhF6kMy5pW from #APress #tns
Wed Mar 07 05:11:11 IST 2012░Louise Corrigan░Technical review position available for #Parse and #Phonegap related @Apress titles: paid,
ongoing work. Email me: louisecorrigan@apress.com
Mon May 11 06:16:04 IST 2009░Hazem Saleh░@RyanBurrell I just contacted Apress team to check the sample zip file. Thanks for your feedback.
Sun Dec 18 20:13:39 IST 2011░cessprin░RT @manthi77: Demain jviens juste en Droit apress fioouuuf jreviens meme pas
```

Figure 6-11. *The sample tweet file delimited by '\001'*

Counting Tweets

In this example we will demonstrate running various Pig commands using the interactive Grunt shell. For Pig scripts having a medium level of complexity, we may want to prepare and run those as Pig scripts, as well. The command to run a Pig script is as follows:

```
Pig -x local myscript.pig
```

Here `myscript.pig` is a compiled Pig script. We can also execute such Pig scripts in embedded mode as follows:

```
// Compile to .class file
javac -cp pig.jar MyScript.java
// Running Pig script as java program in embeddeded mode
java -cp:pig.jar:. MyScript
```

In this exercise, we will explore Apache Pig for running the MapReduce program for total tweet count and counting tweets for a specific `screen_name`.

1. First load tweets using `PigStorage`:

   ```
   tweets = LOAD '/home/vivek/tweets' USING PigStorage('\ua001') as (date:chararray,screen_
   name:chararray,body:chararray);
   ```

2. Let's filter tweets for the screen name The News Selector.

    ```
    name = FILTER tweets by screen_name matches 'The News Selector';
    ```

3. Let's group name for all fields using the GROUP command:

    ```
    namegroup = group name ALL;
    ```

4. To count tweets for namegroup, run the FOREACH command with the COUNT function:

    ```
    tweetCount = FOREACH namegroup generate  COUNT(name);
    dump tweetCount;
    ```

Figure 6-12 shows the output of dumping tweetCount onto the console. The tweet count for screen name The News Selector is 6680.

```
2014-06-15 00:32:17,549 [main] INFO  org.apache.pig.backend.hadoop.executionengine.mapReduceLayer.MapReduceLauncher - Success!
2014-06-15 00:32:17,549 [main] WARN  org.apache.pig.data.SchemaTupleBackend - SchemaTupleBackend has already been initialized
2014-06-15 00:32:17,551 [main] INFO  org.apache.hadoop.mapreduce.lib.input.FileInputFormat - Total input paths to process : 1
2014-06-15 00:32:17,551 [main] INFO  org.apache.pig.backend.hadoop.executionengine.util.MapRedUtil - Total input paths to process : 1
(6680) _
```

Figure 6-12. *Output of dumping tweetCount on console*

5. To store output in a file, we can use the CSVExcelStorage function. To do that, we need to register it first with the Pig registry:

    ```
    register '$PIG_HOME/contrib/piggybank/java/piggybank.jar' ;
    define CSVExcelStorage org.apache.pig.piggybank.storage.CSVExcelStorage();
    ```

Upon running these commands over the Grunt shell, the function will be registered and ready for use.

6. Continuing the same exercise, we can also store the total tweet count as follows:

    ```
    totalGroup = group tweets ALL;
    totalCount = foreach totalGroup generate COUNT(tweets);
    store totalCount into 'totalcount' using CSVExcelStorage(',','YES_MULTILINE');
    ```

A folder named tweetcount will be created in the PIG_HOME directory, which will contain a file with a name like part-r-00000 with the total tweet count.

Until now we have explored Pig for running MapReduce jobs over the local file system. Now let's try to run Pig MapReduce scripts over the Cassandra file system.

It is important to note that we can create complex Pig scripts which may end up running multiple MapReduce jobs. One problem with such Pig scripts is running those jobs in sequence and losing parallel programming. Also you must have noticed the intermediate outputs like loading tweets generated during the running of Pig scripts.

Pig with Cassandra

Cassandra and Pig integration is fairly easy. As mentioned above, to transform Pig Latin scripts into MapReduce over Cassandra requires Cassandra-specific storage functions and connection settings. By default Apache Cassandra comes up with the built-in function support for Pig integration under the package org.apache.cassandra.hadoop.pig.

In this section we will use the CQL-based storage function `CqlStorage` for exercises. Readers may also run the same exercise using `CassandraStorage`, which is primarily for column families created in a non-CQL way. For more details on CQL versus Thrift, please refer to Chapter 1.

The first step is to configure Pig for Cassandra-specific settings:

```
# cassandra daemon host,
export PIG_INITIAL_ADDRESS=localhost
/# thrift rpc port
export PIG_RPC_PORT=9160
#configured partitioner
export PIG_PARTITIONER=org.apache.cassandra.dht.Murmur3Partitioner

#Add thrift library to pig's classpath.
export PIG_CLASSPATH=/home/vivek/software/apache-cassandra-2.0.4/lib/libthrift-0.9.1.jar
```

Data Import

In this section, we will create a Cassandra table to store tweets and use Apache Pig to load tweets in Cassandra.

1. First let's create a keyspace and table in Cassandra as follows:

   ```
   create keyspace twitter with replication = {'class':'SimpleStrategy',
   'replication_factor':1};
   use twitter;
   create table twitterdata(screen_name text primary key, tweetdate text, body text);
   ```

2. Start Pig in local mode and load tweets:

   ```
   tweets = LOAD '/home/vivek/tweets' USING PigStorage('\ua001') as
   (date:chararray,screen_name:chararray,body:chararray);
   ```

You may need to change the directory path as per your settings.

3. Register `apache-cassandra-2.0.4.jar`:

   ```
   register '$CASSANDRA_HOME/lib/apache-cassandra-2.0.4.jar';
   define CqlStorage org.apache.cassandra.hadoop.pig.CqlStorage();
   ```

4. Generate a tuple using TOTUPLE:

   ```
   data_to = FOREACH tweets GENERATE TOTUPLE(TOTUPLE('screen_name',screen_name)),
   TOTUPLE(TOTUPLE('tweetdate',date), body);
   ```

5. Finally, store the generated tuples in Cassandra using the `CqlStorage` function:

   ```
   STORE data_to INTO 'cql://twitter/twitterdata?output_query=update twitterdata set
   tweetdate %3D%3F,body %3D%3F' USING CqlStorage();
   ```

6. At last, connect to the cql shell and verify the loaded data (see Figure 6-13):

   ```
   Select * from twitterdata;
   ```

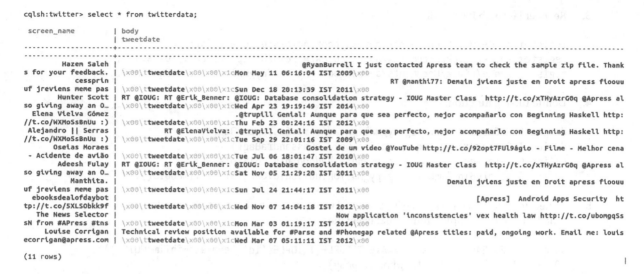

Figure 6-13. *The console depicts the output of data persisted in the twitterdata table*

You must have noticed that total tweet count is 10,020, but still twitterdata in Cassandra got populated with 11 rows. The reason is that multiple tweets are present for the same screen name, and it is defined as the primary key in the twitterdata table.

In the real world, it is certainly possible to have a nonunique field in data files and the user may want Cassandra to take care of the unique key part. We can solve this problem by defining an implicit primary key of type timeuuid and populate it external to original data file. In the next exercise, we will try to solve this problem by introducing a unique primary key.

Loading Sata with timeuuid

Pig Latin does not have any direct support for Cassandra data types such as timeuuid and uuid. But there is an open source project called Pygmalion https://github.com/jeromatron/pygmalion, which provides a number of Pig utilities for Cassandra.

You can git clone or download the zip (https://github.com/jeromatron/pygmalion/archive/master.zip) and build it on your local box. Alternatively, you can find pygmalion-1.1.0-SNAPSHOT.jar in the jars folder from the downloads for this book.

1. First let's create the keyspace and table in Cassandra as follows:

```
create keyspace twitter with replication = {'class':'SimpleStrategy',
'replication_factor':1};
use twitter;
create table twitterdata(id timeuuid primary key, screen_name text, tweetDate text,
body text);
```

2. Load tweets using PigStorage:

```
tweets = LOAD '/home/vivek/tweets' USING PigStorage('\ua001') as
(date:chararray,screen_name:chararray,body:chararray);
```

3. Register jars with Pig registry:

```
register /home/vivek/Documents/apress_book/Apress/uuid-3.2.jar;
register /home/vivek/Documents/apress_book/Apress/hector-core-0.7.0-28.jar;
register /home/vivek/Documents/apress_book/Apress/pygmalion-1.1.0-SNAPSHOT.jar;
```

4. Define the Pig function:

```
define FromCassandraBag org.pygmalion.udf.FromCassandraBag();
define ToCassandraBag org.pygmalion.udf.ToCassandraBag();
define CqlStorage org.apache.cassandra.hadoop.pig.CqlStorage();
define GenerateBinTimeUUID org.pygmalion.udf.uuid.GenerateBinTimeUUID();
```

5. Generate a tuple to have an ID with `timeuuid` values using the `GenerateBinTimeUUID` function and other tuples from actual tweet files:

```
data_to = FOREACH tweets GENERATE TOTUPLE(TOTUPLE('id',GenerateBinTimeUUID())),
TOTUPLE(TOTUPLE('tweetdate',date), body);
```

6. Finally, load this data in Cassandra:

```
STORE data_to INTO 'cql://twitter/twitterdata?output_query=update twitterdata set
tweetdate %3D%3F,body %3D%3F' USING CqlStorage();
```

7. Now you may explore the `twitterdata` column family for inserted data:

```
Select * from twitterdata;
Select count(*) from twitterdata;
```

Up until this point, we have explored various ways to load and run MapReduce programs over Cassandra using Apache Pig. Apache Pig comes in very handy for developers to quickly write Pig Latin scripts to execute MapReduce programs instead of writing lengthy native MapReduce programs.

In next section, we will explore running MapReduce analytics over Cassandra in an SQL manner, which is more commonly used.

Apache Hive

Apache Hive is a platform to provide data analytics support over a very large volume of data stored over HDFS. Hive comes up with various features like built-in UDTF (user-defined table functions), UDAF (user-defined aggregation function), analytics over compressed data, and most importantly Hive Query Language (Hive QL). We will discuss these functions in upcoming sections.

Initially Hive was developed as part of Facebook's research initiatives and later it went on to become an Apache TLP (Top Level Project).

In this section, we will discuss the Hive setup, its execution modes, and integration with Cassandra.

Setup and Configuration

To set up Hive in MapReduce mode, we need to configure Hadoop installation for a few steps. Please note, you may refer to Chapter 5 for more details about MapReduce and Hadoop.

1.   With Hadoop, make sure that JAVA_HOME is configured properly. You may configure
     it by modifying $HADOOP_HOME/conf/hadoop-env.sh and add JAVA_HOME to point jdk
     installation directory:

     ```
     export JAVA_HOME=/usr/lib/jvm/java-7-oracle
     ```

2.   Start the Hadoop installation by running

     ```
     $HADOOP_HOME/bin/start-all.sh
     ```

3.   Please make sure that the name node is properly formatted. Please refer to Chapter 5 for
     more details. If you're getting error such as

     ```
     connect to host localhost port 22: Connection refused
     ```

     you need to install an SSH server on your box. For Ubuntu boxes, you can install it by
     running sudo apt-get install openssh-server.

4.   After this, please verify that all Hadoop processes are running properly (Figure 6-14).

```
vivek@vivek-Vostro-3560:~/software/hadoop-1.1.1$ jps
24200 DataNode
24474 SecondaryNameNode
24922 Jps
23949 NameNode
24568 JobTracker
24872 TaskTracker
```

Figure 6-14. *Shows all Hadoop process are running*

Next, you need to set up and run Hive:

1.   First, download the latest jar from http://apache.mirrors.tds.net/hive/.

2.   Extract the tarball in a local folder and set that local folder as HIVE_HOME.

3.   Set HADOOP_HOME in $HIVE_HOME/bin/hive.sh file.

4.   When this configuration of Hive is complete, we can start the Hive shell by running
     $HIVE_HOME/bin/hive.sh

Understanding UDF, UDAF, and UDTF

Hive comes with built-in user-defined functions (UDF), user-defined aggregate functions (UDAF), and user-defined
table functions (UDTF). Using the Hive shell we can fetch a list of available functions and also describe them:

```
SHOW FUNCTIONS;
DESCRIBE FUNCTION <function_name>;
DESCRIBE FUNCTION EXTENDED <function_name>;
```

The UDFs built in to Hive include functions like round(), pow(), and rand(). And there are built-in collection functions such as mapkeys and map_values to return unordered lists of keys and values respectively. For more about UDFs, refer to https://cwiki.apache.org/confluence/display/Hive/LanguageManual+UDF#LanguageManualUDF-Built-inFunctions.

Among the built-in UDAFs supported by Hive are functions such as count, min, max, and percentile. For a detailed list of supported aggregate functions, you can also refer to https://cwiki.apache.org/confluence/display/Hive/LanguageManual+UDF#LanguageManualUDF-Built-inAggregateFunctions(UDAF).

Also, there are the table-generating UDTFs that operate over multiple rows and at the table level. For example, the explode function generates a row for each array element. Further details and their usage are available at https://cwiki.apache.org/confluence/display/Hive/LanguageManual+UDF#LanguageManualUDF-Built-inTable-GeneratingFunctions(UDTF).

Hive Tables

Hive provides the mechanism for creating two types of tables:

- Hive-managed table
- Hive external table

With Hive-managed tables, Hive is responsible for managing the table's metadata and actual data. Dropping such data would result in data drop as well. Whereas with an external table it will not be managed by Hive, and any drop table activity will not result in data drop. We will explore more about these types of tables in coming exercises.

Local FS Data Loading

In this section, we will discuss more about Hive DDL/DML operations. In this first exercise, we will create a simple schema in Hive and load data from the local file system.

1. First, let's create a database:

    ```
    create database employee_store;
    ```

2. Next, create a table as follows:

    ```
    use employee_store;
    create table employee(person_id string, fname string) row format delimited fields
    terminated by ',';
    ```

Here, with a table DDL operation we have defined that fields are delimited by ', '.

3. We can load data from the local file system as follows:

    ```
    load data local inpath '/home/vivek/Documents/apress_book/Apress/datafiles/
    person_store' overwrite into table employee ;
    ```

4. Finally, we can explore the inserted data (see Figure 6-15):

    ```
    select * from employee;
    ```

```
hive> select * from employee;
OK
p_1      vivek
p_2      Rita
p_3      John
p_4      Apress
Time taken: 0.072 seconds
```

Figure 6-15. *Fetched rows from the employee table*

Let's talk about what a Hive, Hadoop, and Cassandra-based architecture would look like. The image in Figure 6-16 depicts how a Hive-powered MapReduce job would run over Cassandra data nodes. Here each Hive query will get converted into a MapReduce job and submitted with Hadoop which in turn would rely on the name node and metadata to identify Cassandra data nodes to run corresponding maps and reduce tasks locally on those data nodes.

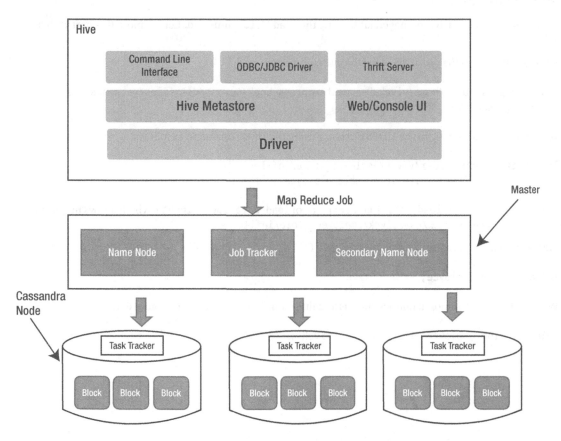

Figure 6-16. *he diagram depicts an architectural representation of running Hive MapReduce over Cassandra data nodes*

One point worth mentioning is running Hive MapReduce jobs over HDFS would be same as shown in Figure 6-17 except data blocks would belong to HDFS in place of Cassandra nodes.

```
hive> load data inpath  '/usr/vivek/employee_store/infile' overwrite into table employee ;
Loading data to table default.employee
Deleted hdfs://localhost:9000/user/hive/warehouse/employee
OK
Time taken: 0.294 seconds
hive> select * from employee;
OK
p_1     vivek
p_2     Rita
p_3     John
p_4     Apress
Time taken: 0.201 seconds
```

Figure 6-17. *Fetching rows from employee table via the Hive shell*

HDFS Data Loading

We can also load from HDFS in a Hive-managed table using the load data command. Let's explore this part with a simple exercise:

1. Let's put the tweets file on HDFS using the put command:

   ```
   bin/hadoop fs -put /home/vivek/Documents/apress_book/Apress/datafiles/person_store
   /usr/vivek/employee_store/infile
   ```

2. We can load data directly to the HDFS directory using the load command as follows:

   ```
   load data inpath  '/usr/vivek/employee_store/infile
                   ' overwrite into table employee ;
   ```

The difference here is that it is load data inpath in place of load data local inpath, which allows mapping from file path to HDFS directory (e.g., /usr/vivek/employee_store/infile).

3. Finally, we can explore inserted data (see Figure 6-17):

   ```
   select * from employee;
   ```

4. We can also use conditional queries such as fetching employees by name (see Figure 6-18):

   ```
   select * from employee where fname = 'vivek';
   ```

```
hive> select * from employee where fname = 'vivek';
Total MapReduce jobs = 1
Launching Job 1 out of 1
Number of reduce tasks is set to 0 since there's no reduce operator
Starting Job = job_201406161219_0001, Tracking URL = http://localhost:50030/jobdetails.jsp?jobid=job_201406161219_0001
Kill Command = /home/vivek/software/hadoop-1.1.1/libexec/../bin/hadoop job  -Dmapred.job.tracker=localhost:9001 -kill job_201406161219_0001
Hadoop job information for Stage-1: number of mappers: 1; number of reducers: 0
2014-06-16 14:19:03,293 Stage-1 map = 0%,  reduce = 0%
2014-06-16 14:19:05,317 Stage-1 map = 100%,  reduce = 0%, Cumulative CPU 0.77 sec
2014-06-16 14:19:06,330 Stage-1 map = 100%,  reduce = 0%, Cumulative CPU 0.77 sec
2014-06-16 14:19:07,344 Stage-1 map = 100%,  reduce = 0%, Cumulative CPU 0.77 sec
2014-06-16 14:19:08,352 Stage-1 map = 100%,  reduce = 0%, Cumulative CPU 0.77 sec
2014-06-16 14:19:09,363 Stage-1 map = 100%,  reduce = 100%, Cumulative CPU 0.77 sec
MapReduce Total cumulative CPU time: 770 msec
Ended Job = job_201406161219_0001
MapReduce Jobs Launched:
Job 0: Map: 1   Cumulative CPU: 0.77 sec   HDFS Read: 250 HDFS Write: 10 SUCCESS
Total MapReduce CPU Time Spent: 770 msec
OK
p_1     vivek
Time taken: 11.572 seconds
```

Figure 6-18. *Fetching rows having the fname vivek*

Upon running such queries, they are transformed into MapReduce jobs unless the conditional column is not a partition key. A query over a partition key can fetch data directly from the data nodes and no MapReduce process will be required as data is already partitioned by partition keys. We can define the partition key while creating the table as follows:

```
CREATE TABLE employee (
    fname   String,
    lname   String,
    emailId    String,
    salary      FLOAT,
)
PARTITIONED BY (location STRING, joining_year INT, joining_month INT, joining_day INT) ;
```

Any conditional query over all parts of a partition key will quickly locate data folders, as data will be automatically arranged into different folders as per the defined partition key.

Hive External Table

We can also create an external Hive table over the HDFS directory. It is recommended to create an external table if it doesn't need to be managed by Hive.

Let's revisit the same tweets example for external table exercises.

1. Let's put the tweets file on HDFS using the put command (see Figure 6-19):

   ```
   bin/hadoop fs -put /home/vivek/Documents/apress_book/Apress/datafiles/person_store
   /usr/vivek/employee_store/infile
   ```

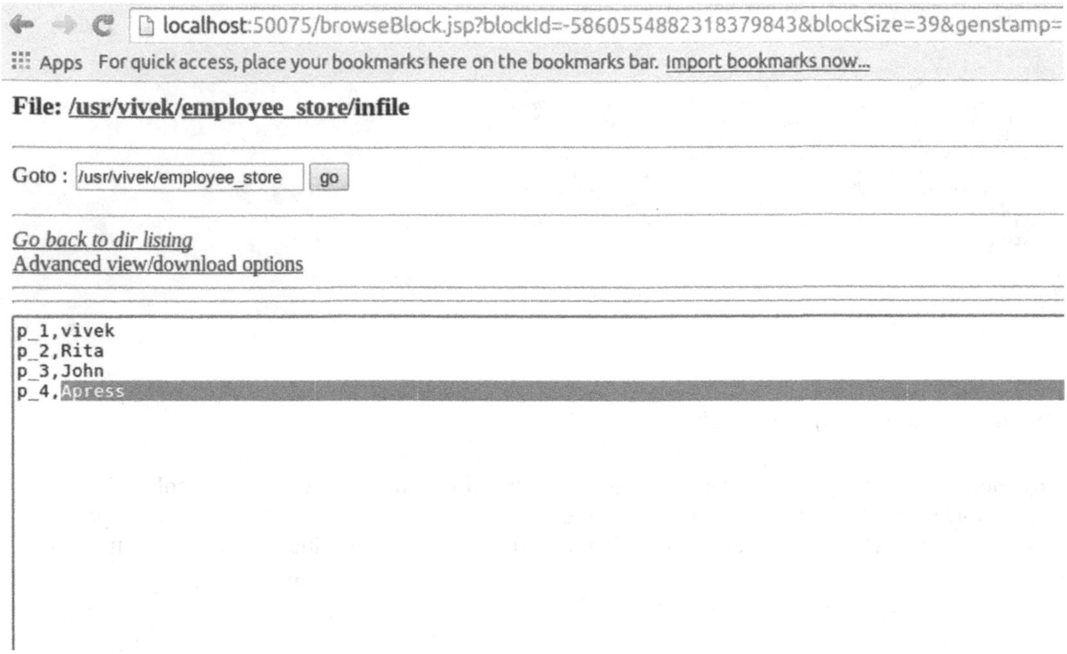

File: /usr/vivek/employee_store/infile

Goto : [/usr/vivek/employee_store] [go]

Go back to dir listing
Advanced view/download options

```
p_1,vivek
p_2,Rita
p_3,John
p_4,Apress
```

Figure 6-19. *A Hive directory with employee records*

2. Then, we can create an external table in Hive over the HDFS location:

```
create external table employee_ext(person_id string, fname string) row format delimited
fields terminated by ',' location 'hdfs://localhost:9000/usr/vivek/employee_store';
```

Here hdfs://localhost:9000 is the value of fs.default.name property.

3. We can explore the inserted data (see Figure 6-20):

```
select * from employee_ext;
```

```
hive> select * from employee_ext;
OK
p_1     vivek
p_2     Rita
p_3     John
p_4     Apress
Time taken: 0.108 seconds
```

Figure 6-20. *Fetching records from the employee_ext external table*

Hive with Cassandra

With external table support it is possible to use Hive for data analytics over Cassandra by using the Cassandra-specific storageHandler implementation. DataStax (www.datastax.com) provides commercial products such as DataStax Enterprise (DSE) which provides seamless integration with tools such as Hive. We will explore more about DSE offering in the "Apache Sqoop" section later in this chapter.

In this section, we will discuss integration of open source Apache Cassandra with Hive. One open source implementation is the Cassandra-specific storage handler that is available at https://github.com/tuplejump/cash. You may git clone or download the zip source and build it locally for jars. Alternatively, you can find these jars under jars folder as a source attachment.

For our example, we will create a table in Cassandra and create external tables over Hive to explore the data inserted via Cassandra and Hive.

1. First, let's create a **twitter** keyspace and **twitterdata** table:

    ```
    create keyspace twitter with replication={'class':'SimpleStrategy',
    'replication_factor':2};
    use twitter;
    create table twitterdata(tweet_id timeuuid primary key, body text, tweeted_by text);
    ```

2. Let's insert a few records with the insert command:

    ```
    insert into twitterdata(tweet_id,body,tweeted_by) values(now(),'my first tweet',
    '@mevivs');

    insert into twitterdata(tweet_id,body,tweeted_by) values(now(),'Cassandra
    book:beginning cassandra development','@apress_team');

    insert into twitterdata(tweet_id,body,tweeted_by) values(now(),
    'Technical review position available for #Parse and #Phonegap related
    @Apress titles: paid','@apress_team');

    insert into twitterdata(tweet_id,body,tweeted_by) values(now(),'Android Apps
    Security http://t.co/5XLSObkk9f','@jhassel');
    ```

3. Let's connect to the Hive shell by configuring the Cassandra storage handler, Thrift, and Cassandra-specific jars:

    ```
    /home/vivek/software/hive-0.9.0/bin/hive --auxpath /home/vivek/source/cash/
    cassandra-handler/target/hive-cassandra-1.2.9.jar:/home/vivek/.m2/repository/org/
    apache/cassandra/cassandra-all/1.2.9/cassandra-all-1.2.9.jar:/home/vivek/software/
    apache-cassandra-2.0.4/lib/libthrift-0.9.1.jar:/home/vivek/.m2/repository/org/
    apache/cassandra/cassandra-thrift/1.2.9/cassandra-thrift-1.2.9.jar
    ```

4. After successfully connecting with the Hive shell, we can create the external table as follows:

    ```
    CREATE EXTERNAL TABLE twitter.twitterdata(tweet_id string, body string,
    tweeted_by string) STORED BY 'org.apache.hadoop.hive.cassandra.cql.CqlStorageHandler'
    WITH SERDEPROPERTIES ("cql.primarykey" = "message_id, author", "comment"="check",
    "read_repair_chance" = "0.2", "dclocal_read_repair_chance" = "0.14", "gc_grace_
    seconds" = "989898", "bloom_filter_fp_chance" = "0.2", "compaction" = "{'class' :
    'LeveledCompactionStrategy'}", "replicate_on_write" = "false", "caching" = "all");
    ```

Please note, support for the Hive-managed Cassandra table is not yet available with the Cassandra storage handler project. Trying to create a Hive-managed table as

```
CREATE TABLE twitter.twitterdata_hive(tweet_id string, body string, tweeted_by string) STORED BY
'org.apache.hadoop.hive.cassandra.cql.CqlStorageHandler' WITH SERDEPROPERTIES ("cql.primarykey" =
"message_id, author", "comment"="check", "read_repair_chance" = "0.2", "dclocal_read_repair_chance"
= "0.14", "gc_grace_seconds" = "989898", "bloom_filter_fp_chance" = "0.2", "compaction" =
"{'class' : 'LeveledCompactionStrategy'}", "replicate_on_write" = "false", "caching" = "all");
```

will result in the error shown in Figure 6-21.

```
hive>
    > CREATE TABLE twitter.twitterdata_hive(tweet_id string, body string, tweeted_by string) STORED BY 'org.apache.hadoop.hive.cassandra.cql.Cq
lStorageHandler' WITH SERDEPROPERTIES ("cql.primarykey" = "message_id, author", "comment"="check", "read_repair_chance" = "0.2", "dclocal_read_
repair_chance" = "0.14", "gc_grace_seconds" = "989898", "bloom_filter_fp_chance" = "0.2", "compaction" = "{'class' : 'LeveledCompactionStrategy
'}", "replicate_on_write" = "false", "caching" = "all");
FAILED: Error in metadata: MetaException(message:Cassandra tables must be external.)
FAILED: Execution Error, return code 1 from org.apache.hadoop.hive.ql.exec.DDLTask
```

Figure 6-21. *shows error while creating internal table over Cassandra in Hive*

Until now, we have explored various ways to perform MapReduce algorithms over Cassandra via Hive and Pig. Another important aspect of data analytics is data migration, where existing RDBMS data needs to be migrated on Cassandra for scalability and performance.

In the next section, we will explore more about data migration.

Data Migration

Legacy applications built using traditional RDBMS and looking to NoSQL databases for scalability and performance may require migrating data from RDBMS to Cassandra. For example, financial applications want to migrate transaction logs, popular websites storing activity logs in RDBMS, and now want to migrate the same to Cassandra.

In the Traditional Way

This is the simplest possible way to export data from an RDBMS such as MySQL in a CSV-format file and load the data in Cassandra. Let's discuss this more with a sample exercise.

1. First, let's create a database and table in MySQL:

   ```
   create schema twitter;
   create table twitterdata(id varchar(20) primary key,screen_name varchar(20),
   body varchar(30));
   ```

2. Insert some sample tweets:

   ```
   insert into twitterdata('1','mevivs','my first tweet');
   insert into twitterdata values('2','jhassel','my first tweet');
   insert into twitterdata values('3','rfernando','my first tweet');
   ```

3. Let's export data from the `twitterdata` table:

```
select * from twitterdata INTO OUTFILE '/tmp/mysql_output.csv' Fields TERMINATED BY ','
ENCLOSED BY '"' LINES TERMINATED BY '\n';
```

4. Next, let's create a table and keyspace in Cassandra:

```
create table ntwitterdata(id text primary key, screen_name text, bodt text);
```

5. Finally, we can copy `twitterdata` from the csv file using the copy command over the cql shell:

```
copy twitterdata from '/tmp/mysql_output.csv';
```

This way, we can perform data migration in the traditional way. Unfortunately it happens in the localized way.

It looks simple, but what about migrating the complete schema containing multiple tables? Would giving the name of the schema would be more than sufficient to establish tunnel and migrate data? Apache Sqoop (SQL to Hadoop) is an answer for these questions.

Apache Sqoop

Apache Sqoop is a tool to transfer data from relational databases to NoSQL databases such as Cassandra or distributed file systems such as Hadoop. Sqoop has been an Apache TLP since 2012. Sqoop comes in very handy when we need to establish a tunnel for data migration between RDBMS and NoSQL (e.g., Cassandra, Hadoop).

We will be using DataStax's DSE for sample Sqoop Cassandra integration in this section. For more details about DSE setup and configuration you can refer to

```
http://www.datastax.com/docs/datastax_enterprise3.1/reference/start_stop_dse
http://www.datastax.com/docs/datastax_enterprise3.0/install/install_rpm_pkg
```

You can also download and extract the tarball in a local folder.

Sqoop with Cassandra

For a Sample Exercise We Will Use the Same Tweets File We Used Previously and Then Finally use DSE Sqoop Support to Migrate From MySQL to Cassandra.

1. First, let's connect to the MySQL client:

```
mysql -u root -p –local-infile
```

2. Next, we will create a keyspace **twitter** and table **twitterdata**:

```
create keyspace twitter with replication={'class':'SimpleStrategy',
'replication_factor':2};
use twitter;
create table twitterdata(tweetdate varchar(50),screen_name varchar(50),
body varchar(300), id int NOT NULL AUTO_INCREMENT, PRIMARY KEY(id));
```

3.   Let's load tweets data in the `twitterdata` table (see Figure 6-22):

```
LOAD DATA LOCAL INFILE '/home/vivek/tweets' INTO TABLE twitterdata FIELDS
TERMINATED BY 0x01 LINES TERMINATED BY '\n';
```

```
mysql> LOAD DATA LOCAL INFILE '/home/vivek/tweets' INTO TABLE twitterdata FIELDS TERMINATED BY 0x01 LINES TERMINATED BY '\n';
Query OK, 10020 rows affected, 10020 warnings (0.17 sec)
Records: 10020  Deleted: 0  Skipped: 0  Warnings: 10020
```

Figure 6-22. *Loading the file from the local file system in the hive table*

4.   Next, start DSE Cassandra with Hadoop:

```
bin/dse cassandra -t
```

5.   After this, run `sqoop import` as follows:

```
bin/dse sqoop import --connect jdbc:mysql://localhost/twitter --username root -P
--table twitterdata --cassandra-keyspace twitter --cassandra-column-family twitterdata
--cassandra-row-key id --cassandra-thrift-host localhost --cassandra-create-schema
```

6.   After successfully importing, let's explore the `twitterdata` column family via `cassandra-cli` (see Figure 6-23):

```
$DSE_HOME/bin/cassandra-cli
```

```
RowKey: 6130
=> (name=body, value=Obama meeting with European allies on Ukraine http://t.co/RtjWYwPpEs from #APress #tns, timestamp=1402878157525)
=> (name=screen_name, value=The News Selector, timestamp=1402878157525)
=> (name=tweetdate, value=Mon Mar 03 01:19:17 IST 2014, timestamp=1402878157525)
-------------------
```

Figure 6-23. *Exploring cassandra-cli for loaded data*

7.   You can explore `twitterdata` table from `cqlsh` (see Figure 6-24):

```
$DSE_HOME/bin/cqlsh
```

```
cqlsh:twitter> describe table twitterdata;

CREATE TABLE twitterdata (
  key text,
  column1 text,
  value text,
  PRIMARY KEY (key, column1)
) WITH COMPACT STORAGE AND
  bloom_filter_fp_chance=0.010000 AND
  caching='KEYS_ONLY' AND
  comment='' AND
  dclocal_read_repair_chance=0.000000 AND
  gc_grace_seconds=864000 AND
  read_repair_chance=0.100000 AND
  replicate_on_write='true' AND
  populate_io_cache_on_flush='false' AND
  compaction={'class': 'SizeTieredCompactionStrategy'} AND
  compression={'sstable_compression': 'SnappyCompressor'};
```

Figure 6-24. *Describing the table twitterdata*

With this, we conclude that Sqoop can easily be configured with Cassandra for data migration purposes. The next chapter will take discussion forward with graph databases.

Summary

The following is a summary of topics discussed in this chapter:

- Configure and run Hive-based queries to integrate and run MapReduce jobs in a SQL-like manner.

- Pig comes very handy if the reader is well-versed with script-based programming and familiar with writing scripts.

- Use Sqoop for migrating data from HDFS/MySQL to Cassandra and vice versa.

The next chapter will take the discussion ahead with new the paradigm, graph-based databases. The chapter will discuss why, how, and when to use graph databases and most importantly how to do so with Cassandra.

■ ■ ■

Titan Graph Databases with Cassandra

So far, we have discussed data modeling in Cassandra, building a large data analytics platform using Hadoop and related technologies. When business requirements require interconnectedness among various objects, an answer is having reference (joins) queries among multiple objects. Everything looks good until we have to deal with many such relations or the data volume becomes really huge.

LinkedIn, Twitter, and Facebook are popular examples where relationships among registered users can be described as a social graph. Another example could be building a credit history and doing risk analysis of granting loans to a group of customers. Definitely with multiple tables and join theory, we can implement a solution but obviously it will not be scalable and performant. That's where idea of using a graph database comes very handy.

In this chapter we will discuss:

- An introduction to graph concepts

- Graph frameworks and databases

- Titan Graph setup and installation

- Graph use cases

Introduction to Graphs

Graph theory can be traced back to 1736. A graph is a data structure that consists of vertices/nodes and edges. A graph can have zero or multiple edges. A graph without edges is also referred to as an empty graph.

A vertex is a graph node that is connected to other graph nodes/vertices via edges. Each edge is an arc or line which connects multiple vertices.

In computer science, a graph can be categorized in many ways. A few of the popular ones include:

- Simple and nonsimple graphs

- Directed and undirected graphs

- Cyclic and acyclic graphs

Simple and Nonsimple Graphs

A simple graph, as the name suggests, is a basic graph having at most one edge between two vertices. It's a graph that has no self loops but multiple nodes (see Figure 7-1a). On the other hand, graphs with self loops and multiple edges are nonsimple graphs (see Figure 7-1b). Here, a self loop is an edge which connects to a vertex itself. A graph having multiple edges between the same nodes is also called a multigraph or parallel graph. The graph shown in Figure 7-1c is a multigraph and a self-loop graph. It contains a self loop on vertex A and has multiple edges between vertex A and B.

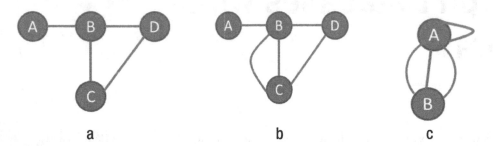

Figure 7-1. *a) On the left, a simple graph; b) a nonsimple graph in the middle; and c) a multigraph with a self loop*

Directed and Undirected Graphs

Graphs with multiple nodes and directed edges are directed graphs (see Figure 7-2a). Undirected graphs are the graphs having edges without direction (see Figure 7-2b).

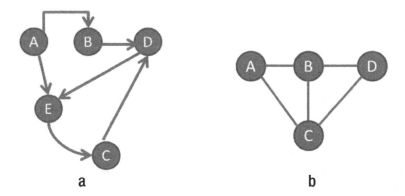

Figure 7-2. *a) On the left, a directed graph with directed edges; b) an undirected graph*

Cyclic and Acyclic Graphs

A graph is said to have a cycle if that traverses a path with at least one edge and starts and ends at the same vertex, whereas a non-cyclic or acyclic graph doesn't contain any cycle. Such graphs can be directed or undirected. Figure 7-3 shows graphical representations of cyclic and acylic graphs.

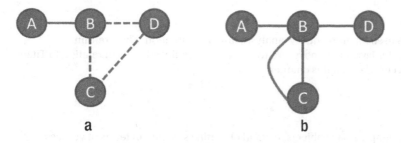

Figure 7-3. *a) The dotted lines show a cycle in a graph; b) an acyclic graph*

Figure 7-3a represents a cyclic graph with edges represented by dotted lines connecting the B, C, and D vertices.

Those are a few basic types of graphs. The next question that may come to mind is whether there are any open source tools, frameworks, or databases specifically built to handle graph-related problems?

Open Source Software for Graphs

Open source software is free, easy to submit bugs and request feature modifications according to our needs, and most importantly cost effective. In recent years, using open source software in the IT industry has become popular, and more organizations prefer open source solutions. A few of the considerations before adopting an open source solution are:

- Should be mature and stable

- Should be in active development and must have community support

- There must be systems in production to validate industry usage

We asked the question of whether there are any tools, frameworks, or databases for solving graph-related problems. Well, let's explore and find out!

Graph Frameworks: TinkerPop

Graph frameworks are used for graph data modelling and visualization. In this section we will discuss TinkerPop (www.tinkerpop.com/) and its feature set. TinkerPop Blueprints is used as a specification by many NoSQL databases (including Titan). Blueprints provides a set of interfaces and implementations for graph data modeling and will be discussed in later in this section.

TinkerPop is an open source graph computing framework with multiple components, frameworks, and command-line tools for handling graph data modeling and visualization. In this section we will discuss them individually.

Pipes

Pipes is a dataflow framework that enables the splitting, merging, filtering, and transformation of data from input to output. Computations are evaluated in a memory-efficient, lazy fashion.

Think of pipes as vertices that are connected by edges, with functions for extraction, transformation, and data computation generally.

Gremlin

Gremlin is a graph traversal language that is used for graph query, analysis, and manipulation. The Gremlin distribution comes with built-in API support for Java and Groovy. We will discuss Gremlin at length in relation to Titan in the "Command-line Tools and Clients" section later in this chapter.

Frames

Frames exposes the elements of a Blueprints graph as Java objects. Instead of writing software in terms of vertices and edges, with Frames, software is written in terms of domain objects and their relationships to each other.

Rexster

Rexster is a multi-faceted graph server that exposes any Blueprints graph through several mechanisms with a general focus on REST. It exposes a graph server via the REST API and RexPro protocol. RexPro is a binary protocol for Rexster that can be used to send Gremlin scripts to a remote Rexster instance. The script is processed on the server and the results serialized and returned to the calling client. It also provides tools for a browser-based interface known as the Dog House and the Rexster console (which will be discussed with the Titan ecosystem).

Furnace

Furnace is a property graph algorithms package. It provides implementations for standard graph analysis algorithms that can be applied to property graphs in a meaningful way. Furnace provides different graph algorithm implementations that are optimized for different graph computing scenarios, such as single-machine graphs and distributed graphs.

■ **Note** Single machine graphs involve graph data over a single node, whereas distributed graphs have data distributed across multiple nodes.

Blueprints

Blueprints, as the name suggests, is a property graph model interface with provided implementations. Databases that implement the Blueprints interface automatically support Blueprints-enabled applications.

Blueprints can be thought of as JDBC (Java DataBase Connectivity) or JPA (Java Persistence API) APIs for graph databases. Most graph databases implement Blueprints. Figure 7-4 shows a representation of how Blueprints can be visualized with the previously mentioned TinkerPop components.

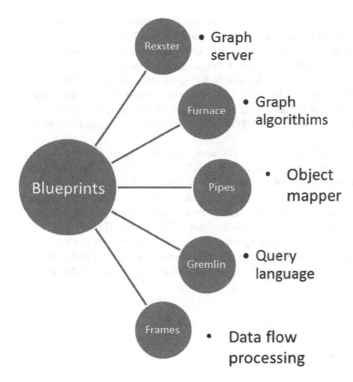

Figure 7-4. *TinkerPop Blueprints implementations*

So with this we have covered TinkerPop framework, its components and other graph-related concepts. The next question is whether there are any graph-based solutions that can be thought of as graph databases. In next section we will answer these questions.

Graph as a Database

In comparison to traditional RDBMS, NoSQL databases are less about schema and more about denormalized forms of data. But graph databases offer the flexibility to define relationships between nodes via edges, and that's why it is easy to understand them in terms of RDBMS concepts. Building graph-like queries with an RDBMS is certainly possible but as discussed previously it will be of very limited use. With non-graph databases, the ability to run graph-based queries for traversal or building graph structures is not supported and could be cumbersome to build. Because of inherent graph data structure support, graph databases will have an edge over traditional RDBMS.

A few differences between RDBMS and graph databases are

- There is no need for index lookup with graph databases, as nodes or vertices are aware of properties they are connected with (e.g., edges) whereas with RDBMS we need to rely on an indexing mechanism.

- Two vertices interconnected via edges can be different in properties and may evolve dynamically, but RDBMS imposes a fixed set of schema.

- With graph databases, the relationship between two vertices is stored at the record level whereas with RDBMS it is defined at the table level.

One thing we need to keep in mind is that the current era of application development is one of using **specific technologies for specific needs**, which is a good fit for building polyglot or hybrid solutions. This means that for cases in which your needs are best served by running graph-like queries and your requirements lend themselves to a faster graph-based model, then the answer is simple: Use graph databases. A graph database uses nodes and edges and their properties to build and store data in a graph structure. Every element in a graph databases has a reference to adjoining nodes, which means no index lookup will be required. Figure 7-4 shows an example of a graph database storing Twitter users as nodes and their followers as edges. Each node contains an fname, id, lname, and role as properties, whereas each edge has a property to denote the date when a user became a follower of the adjoining node (i.e., user).

Figure 7-5 shows a Twitter connection and follower graph for users mevivs, chris_n, apress_team, and melissa_m. Here the vertex apress_team is being followed by the mevivs and melissa_m vertices. On the other hand, a transitive relation/traversal exists between chris_n, who is following mevivs, who follows apress_team, and the apress_team follows chris_n. In the "Gremlin Shell" and "Use Cases" sections, we will refer to the same Twitter example to explore command-line tools and Java APIs in sample exercises. When considering such transitive graph queries, one thing worth discussing is that the ways graph databases handle such queries is different than SQL queries. Handling of such transitive queries with SQL is not straightforward and would require performing complex joins, unions, or intersections. But handling such transitive queries with graph databases is much easier and requires **just following the edges** for incoming and outgoing data queries using the edge's properties. The "Use Cases" section will discuss these graph traversals.

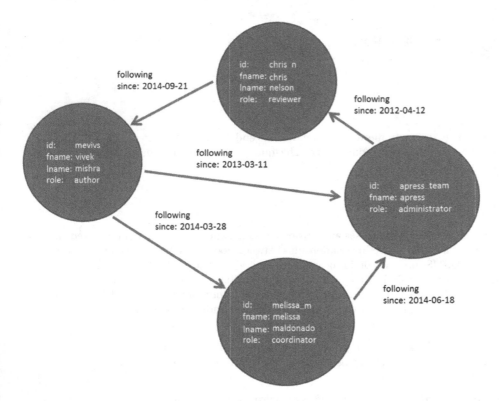

Figure 7-5. A graph database storing Twitter users, their properties, and followers

Let's talk about available graph databases. A few of the popular graph databases are

- Neo4J
- OrientDB
- Objectivity's InfiniteGraph
- Titan

Because the intent of this chapter is to discuss Cassandra and Titan graph databases at length, we will discuss only important features provided by the other databases mentioned here.

Neo4J

Neo4J (www.neo4j.org/) is an open source database licensed under GPU (General Public Usage). It was developed by Neo Technology Inc. It stores data in the form of nodes that are connected by directed edges. A few important features provided include:

- Scalable and highly available data
- Fll ACID transaction support
- REST-based access
- Support for Cypher and Gremlin graph query language

OrientDB

OrientDB (www.orientechnologies.com/orientdb/) is an Apache 2 licensed NoSQL graph database. It is managed and developed by Luca Garulli of Orient Technologies. A few important features provided by OrientDB are:

- REST-based access
- Full ACID transaction support
- A SQL-like interface for query support

InfiniteGraph

InfiniteGraph (www.infinitegraph.com/) is an enterprise distributed graph database built by Objectivity Inc. A few important features supported by InfiniteGraph are

- Support for concurrency and consistency
- Full ACID transaction support
- A visualization tool

Titan

Titan (`thinkaurelius.github.io/titan/`) is an Apache licensed scalable graph database built to store a large amount data in the form of nodes and edges. It supports Cassandra as backend storage. A few of its important features are

- Full text search and geospatial query support via Lucene/ES

- ACID support

- Eventual and intermediate consistency

- Support for multiple databases which can be good for polyglot graph-based applications

- Support for Gremlin and cypher

Titan and its various components will be discussed in the coming sections.

Titan Graph Databases

Titan is a transactional graph database that allows thousands of concurrent users to execute complex graph traversal queries in real time. It also provides support for graph data analytics, reporting, and ETL support via Hadoop integration. It also comes with built-in support of Elasticsearch and Lucene for geospatial queries and full text search. It also provides native support for a Blueprint TinkerPop graph stack.

Figure 7-6 shows a graphical representation of the Titan ecosystem.

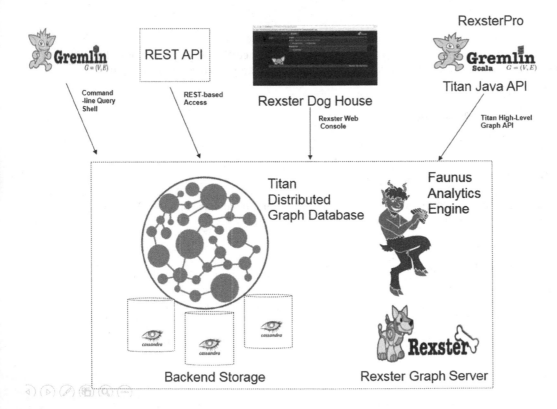

Figure 7-6. *The Titan ecosystem*

Basic Concepts

In this section, we'll introduce some basic concepts that are important for understanding Titan graph databases.

Vertex-Centric Indices

A vertex-centric index is specific to a vertex. Most traversal among referencing or non-referencing vertices would be via edges or their properties. Indexing such properties or edge labels can avoid performance overhead. Such indices can also be referred to as local indices. The purpose of vertex-centric indices is to sort and index adjoining edges of a vertex based on an edge's properties and labels.

Titan also provides support for Elasticsearch, which can be run in standalone or embedded mode with Titan. With Elasticsearch it is also possible to perform full text search, executing geospatial queries, and numeric range queries. Elasticsearch allows us to query over nonindexed properties, as well.

Edge Compression

With edge compression, Titan can store compressed metadata and keep memory usage low. It also can store all edge labels and properties within the same data block for faster retrieval.

Graph Partitioning

This is where the underlying database matters the most. In Cassandra, we know that data for one particular row would be stored on one Cassandra node. Titan understands that with keys sorted by vertex ID, it can effectively partition and distribute graph data across Cassandra nodes. The vertex ID is a 64-bit unique value.

By default Titan adopts a random partitioning strategy to randomly assign vertices to nodes. Random partitions are efficient and keep the cluster balanced, but in the case of multiple backend storage options adopted for storing graphs, it would lead to performance issues and require cross-database communication across the instances. With Titan we can also configure explicit partitioning. With explicit partitioning we can control and store traversed subgraphs on same node.

We can enable an explicit partition in Titan as follows:

```
cluster.partition = true
cluster.max-partitions = 16
ids.flush = false
```

Here `max-partitions` is the maximum number of partitions per cluster. Also we need to disable flushing IDs as well.

When using Cassandra as the storage backend option, it must be configured with the `ByteOrderedPartitioner`.

Titan stores vertices and adjoining edges and properties as a collection. This mechanism is called **adjacency list format.** With this mechanism, edges connecting to a vertex and its properties would be stored together in a collocated way. Each row will contain a vertex ID as a key and adjoining edges and properties as cells. Data representation in this format is common across all supported databases.

Figures 7-7 and 7-8 show the Titan data layout and edge storage mechanisms. (The images are from the Titan wiki page at `https://github.com/thinkaurelius/titan/wiki/Titan-Data-Model` and reused under the Apache License, Version 2.0, `www.apache.org/licenses/LICENSE-2.0.`)

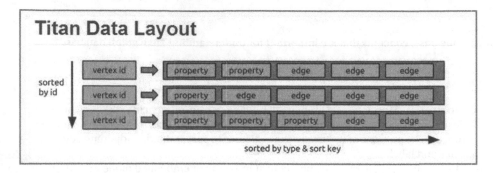

Figure 7-7. Titan's data layout mechanism

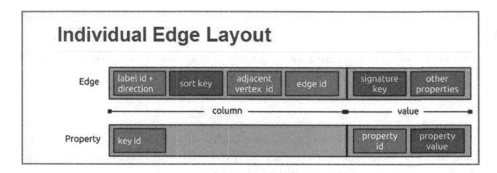

Figure 7-8. Titan's edge storage mechanism

The underlying datastore will store each vertex along with adjoining edges and properties as a single row. Also these rows will be sorted by vertex id and cells will be sorted by property and edge key. Dynamic cells can be added at run time. Data collocation is very important, that's why storing the vertex and adjoining edges as a single row would help to achieve high availability.

Backend Stores

Titan's storage architecture is totally decoupled and supports multiple NoSQL databases, which are

- Cassandra
- HBase
- BerkeleyDB
- Persistit

Support for multiple NoSQL data stores allows adopting the right one based on application requirements. In other words, you can select specific technology for specific needs. Based on the CAP theorem we may opt for any one of the supported databases.

With Titan we can configure backend storage on the fly using the **storage.backend** option. Examples in this chapter will cover how to use this option with Cassandra.

Transaction Handling

Titan is a transactional graph database; hence every read/write operation would happen in a transaction boundary.
The following code snippet shows Titan wrap the vertex mevivs in a transaction boundary and commit it:

```
TitanGraph g = TitanFactory.open("/home/vivek/Titan");
Vertex mevivs = g.addVertex(null); //Implicitly wraps within transaction
mevivs.setProperty("fname", "vivek");
g.commit(); //Commits transaction
```

In cases with very large volume and a polyglot nature, permanent or temporary failures may happen. Here
temporary failures are situations such as network failure, nodes not responding, and similar scenarios. In such
scenarios we can configure the retry delay property with Titan like this:

```
Configuration conf = new BaseConfiguration();
conf.setProperty("storage.attempt-wait ",250); // time in milliseconds
```

Such temporary failure can be handled with retries, but permanent failure, like hardware failure, would require
the user to explicitly handle TitanException:

```
Try
{
    TitanGraph g = TitanFactory.open("/home/vivek/Titan");
    Vertex mevivs = g.addVertex(null); //Implicitly wraps within transaction
    mevivs.setProperty("fname", "vivek");
    g.commit(); //Commits transaction
} catch (TitanException e) {
    //configure explicit retry or fast-fail scenarios.
}
```

These are the Titan basic concepts and its architecture. Next, we will cover setup and installation of
Titan Graph database.

Setup and Installation

We can download the latest Titan distribution from https://github.com/thinkaurelius/titan/wiki/Downloads.
The latest version at the time of writing is 0.5.0. After downloading the distribution, extract it to a local folder. We will
be referring to the TITAN_HOME variable at many places in this chapter. We can set it as follows:

```
export TITAN_HOME=/home/vivek/software/titan
```

Command-line Tools and Clients

With setup and installation in place, the first question that comes to mind is whether there are any command-line
clients? Like the CQL shell for Cassandra, is there any option available with Titan for server-side scripting and quick
analysis? Gremlin shell and Rexster are two command-line options we will be exploring in this section.

Gremlin Shell

Titan provides support for Gremlin shell for graph traversal and mutation using the Gremlin query language. It's a functional language. Each step outputs an object, and with "." (dot), we can access associated functions with it. For example,

```
gremlin> conf = new BaseConfiguration() // step 1
==>org.apache.commons.configuration.BaseConfiguration@2d3c117a

gremlin> conf.setProperty("storage.backend", "cassandrathrift") // step 2
```

Here conf is an object of Configuration created in step 1 whose setProperty function has been invoked in step 2.

Let's discuss Gremlin with an exercise. In this recipe we will be using the same Twitter example, where users' tweets will be a graph's vertices and the relationship between a user and its tweets and followers will be edges. In this example we will be using Cassandra as the backend storage option. You can opt for running a standalone Cassandra server; otherwise, by default, it would start and connect with an embedded one.

1. First we need to connect with Gremlin as in Figure 7-9.

```
vivek@vivek-Vostro-3560:~/software/titan-all-0.4.4$ bin/gremlin.sh

         \,,,/
         (o o)
-----oOOo-(_)-oOOo-----
gremlin> ▊
```

Figure 7-9. *Connected to the Gremlin shell*

2. Next, we need to initialize a configuration object and set a few Cassandra-specific properties:

```
gremlin> conf = new BaseConfiguration()
==>org.apache.commons.configuration.BaseConfiguration@2d3c117a
gremlin> conf.setProperty("storage.backend", "cassandrathrift")
==>null
gremlin> conf.setProperty("storage.hostname", "localhost")
==>null
gremlin> conf.setProperty("storage.port", "9160")
==>null
gremlin> conf.setProperty("storage.keyspace", "twitter")
==>null
```

3. Next, get an object of Titan graph:

```
gremlin> graph = TitanFactory.open(conf)
==>titangraph[cassandrathrift:localhost]
```

4. Let's make few vertex keys and edge labels:

```
gremlin> graph.makeKey("fname").
dataType(String.class).indexed(Vertex.class).make()
==>fname
gremlin> graph.makeKey("lname").
dataType(String.class).indexed(Vertex.class).make()
==>lname
gremlin> graph.makeKey("twitter_tag").
dataType(String.class).indexed(Vertex.class).make()
==>twitter_tag
gremlin> graph.makeKey("tweeted_at").
dataType(String.class).indexed(Vertex.class).make()
==>tweeted_at
gremlin> graph.makeLabel("has_tweeted").make()
==>has_tweeted
gremlin> tweet.setProperty("body", "Working on Cassandra book for apress")
==>null
gremlin> tweet.setProperty("tweeted_at", "2014-09-21")
==>null
```

Here fname, lname, and twitter_tag are vertex keys and the label has_tweeted will be used for edges in the next step.

5. Let's create vertices for user and tweet:

```
gremlin> vivs = graph.addVertex(null)
==>v[4]
gremlin> vivs.setProperty("fname", "vivek")
==>null
gremlin> vivs.setProperty("lname", "mishra")
==>null
gremlin> vivs.setProperty("twitter_tag", "mevivs")
==>null
gremlin> graph.V("fname","vivek")
==>v[4]
gremlin> tweet = graph.addVertex(null)
==>v[8]
```

6. Next, add an edge between these two vertices:

```
gremlin> graph.addEdge(null, vivs, tweet, "has_tweeted")
==>e[2V-4-1E][4-has_tweeted->8]
```

7. Let's add apress_team as a user and establish and define "vivek follows apress_team" relationship edge:

```
gremlin> apress = graph.addVertex(null)
==>v[12]
gremlin> apress.setProperty("fname", "apress")
==>null
```

```
gremlin> apress.setProperty("twitter_tag", "apress_team")
==>null
gremlin> graph.addEdge(null, vivs, apress, "following")
==>e[3r-4-22][4-following->12]
```

8. We can find a vertex by its key as follows:

```
gremlin> vivek = graph.V('fname','vivek').next()
==>v[4]
gremlin> vivek.map()
==>twitter_tag=mevivs
==>lname=mishra
==>fname=vivek
```

9. We can also fetch all outgoing edges from vertex vivek having the relationship has:

```
gremlin> outVertex = vivek.out('has').next()
==>v[8]
gremlin> outVertex.map()
==>body=Working on Cassandra book for apress
==>tweeted_at=2014-09-21
```

The preceding recipe demonstrates a way to populate and traverse through a Twitter graph application using Gremlin query language.

Let's discuss Rexster Rest API, the Dog House, and Titan Server.

Rexster: Server, Rest API, and the Dog House

As discussed in the TinkerPop section, using the REST API and web console, we can visualize and manage any Titan graph. In a previous recipe we discussed downloading and setting up the Titan distribution on a local box. To start Titan Server, embedded Elasticsearch, and Cassandra, we need to run

```
TITAN_HOME/bin/titan.sh
```

Next, to connect with the REST API and the Dog House we need to execute

```
TITAN_HOME/bin/rexster-console.sh
```

This will start Elasticsearch and connect with Elasticsearch transport at port 9300 and will get Rexster running at port 8184. The REST API and the Dog House console would get started on localhost:8182 port (see Figure 7-10).

```
vivek@vivek-Vostro-3560:~/software/titan-0.5.0-hadoop1/bin$ ./titan.sh start
Forking Cassandra...
Running `nodetool statusthrift`.... OK (returned exit status 0 and printed string "running").
Forking Elasticsearch...
Connecting to Elasticsearch (127.0.0.1:9300)... OK (connected to 127.0.0.1:9300).
Forking Titan + Rexster...
Connecting to Titan + Rexster (127.0.0.1:8184).......... OK (connected to 127.0.0.1:8184).
Run rexster-console.sh to connect.
vivek@vivek-Vostro-3560:~/software/titan-0.5.0-hadoop1/bin$ █
```

Figure 7-10. *Starting Rexster server*

Rexster Dog House

Figure 7-11 shows the Dog House console with tabs for the Dashboard, the option to browse edges and vertices, and the built-in Gremlin command-line shell.

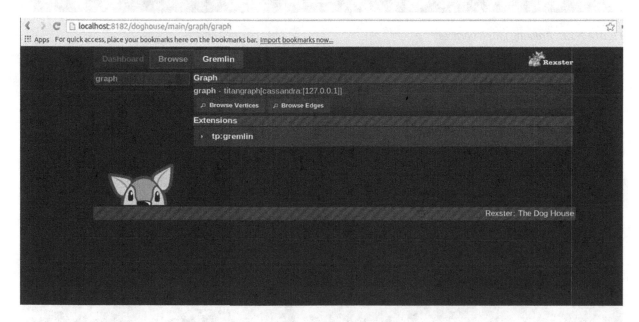

Figure 7-11. *Rexster Dog House web console*

The built-in Gremlin command-line client allows you to run graph mutation and traversal queries (discussed previously in this chapter).

Let's explore the graph stored in the Gremlin recipe using Gremlin query language. We can analyze vertices using the Browse Vertices option as shown in Figure 7-12.

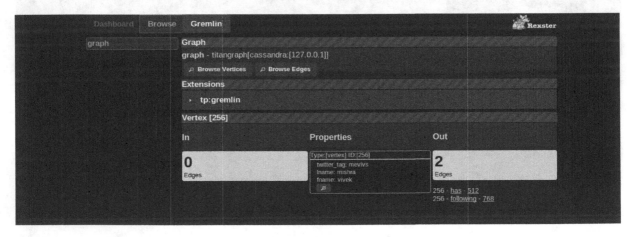

Figure 7-12. *Browing graph vertices*

We can further drill down to properties of a specific vertex, as well (see Figure 7-13). The figure shows the properties and incoming and outgoing from vertex "vivek". It has two outgoing edges "has" and "following" to tweet and apress_team vertex.

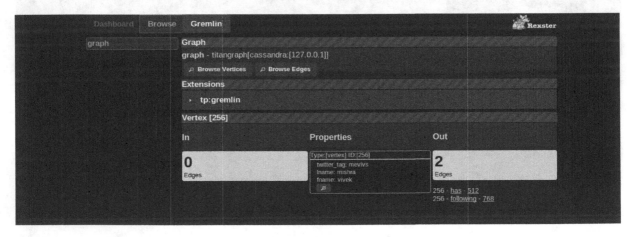

Figure 7-13. *Exploring properties and relationships of vertex "vivek"*

In the same way we can also explore edges and their properties (see Figures 7-14 and 7-15).

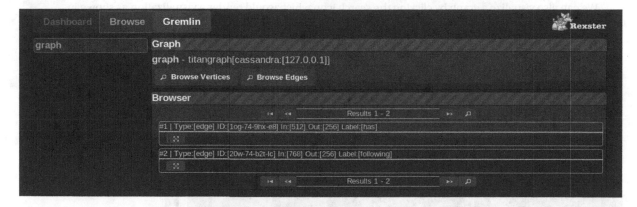

Figure 7-14. *Exploring the edges with the labels "has" and "following"*

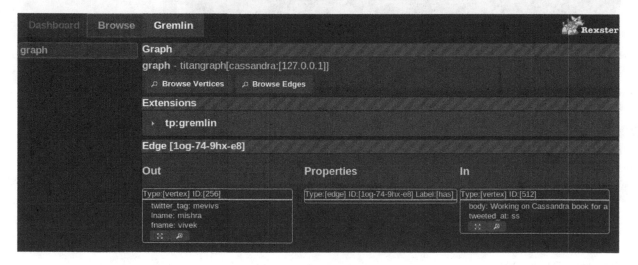

Figure 7-15. *Exploring the edge with the label "has"*

Figure 7-15 shows the properties and connected incoming and outgoing vertices of edge "has". The vertices mevivs and tweet are connected via the edge "has" which in lay terms means "Vivek has tweeted a tweet!"

We can also visualize a graph by clicking the search icon which would render a visualization as shown in Figure 7-16.

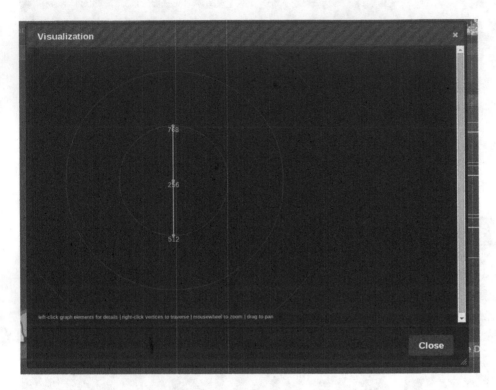

Figure 7-16. *Visualization of vertices and edges*

The figure depicts a graphical representation of three vertices and two edges.

Rexster REST API

We can also query a Titan graph database using the REST API! For example, to get a list of all vertices we simply need to hit a request like this:

```
localhost:8182/graphs/$graph_name/vertices
```

For example, we can get a list of all vertices of the twitter "graph" as shown in Figure 7-17.

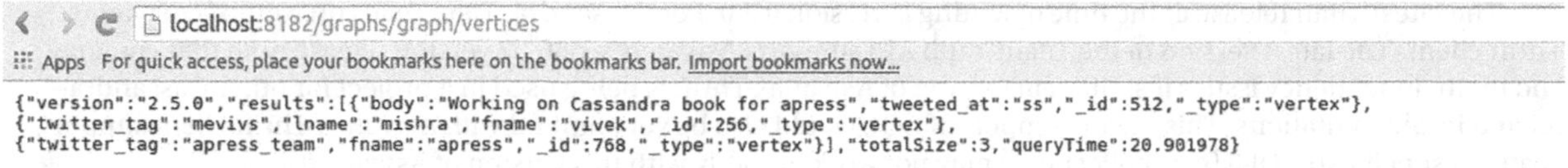

localhost:8182/graphs/graph/vertices

Apps   For quick access, place your bookmarks here on the bookmarks bar. Import bookmarks now...

{"version":"2.5.0","results":[{"body":"Working on Cassandra book for apress","tweeted_at":"ss","_id":512,"_type":"vertex"},
{"twitter_tag":"mevivs","lname":"mishra","fname":"vivek","_id":256,"_type":"vertex"},
{"twitter_tag":"apress_team","fname":"apress","_id":768,"_type":"vertex"}],"totalSize":3,"queryTime":20.901978}

Figure 7-17. *Returning a list of all vertices using the REST API*

We can also query for a specific vertex by its ID (see Figure 7-18).

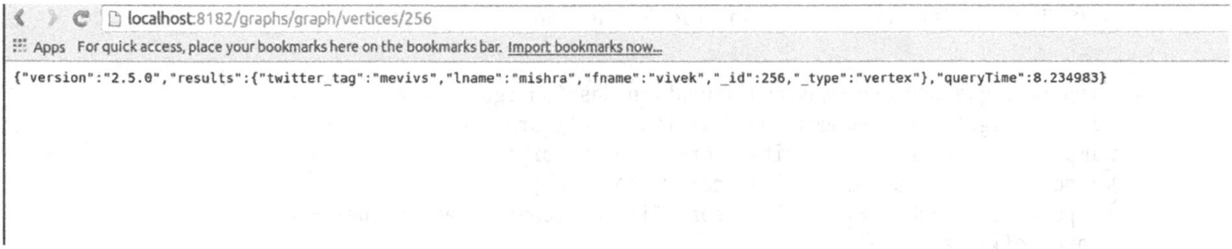

localhost:8182/graphs/graph/vertices/256

Apps   For quick access, place your bookmarks here on the bookmarks bar. Import bookmarks now...

{"version":"2.5.0","results":{"twitter_tag":"mevivs","lname":"mishra","fname":"vivek","_id":256,"_type":"vertex"},"queryTime":8.234983}

Figure 7-18. *Get a vertex using the ID 256*

A complete list of all supported REST methods can be found at `https://github.com/tinkerpop/rexster/wiki/Basic-REST-API`.

Titan with Cassandra

In this section we will discuss the Titan implementation using Cassandra as a storage option. This section will discuss how to use the Titan Java API with Cassandra and perform use cases such as reading and writing to graphs.

Titan Java API

Titan is an implementation of the Blueprints graph interface. Titan provides Java- and Groovy-based implementations to access Titan.

The Titan Java API setup is fairly easy. For backend storage it relies on other databases, so to start using Titan we just need to add

```
<dependency>
  <groupId>com.thinkaurelius.titan</groupId>
  <artifactId>titan-all</artifactId>
  <version>0.4.4</version>
</dependency>
```

The latest Titan release at the time of writing is version 0.5.0. For Cassandra, Titan relies on Netflix's Astyanax Thrift client. The latest version of the TitanGraph API supports Astyanax's 1.56.37 version. Please note that you may end up in dependency issues if a different version of Astyanax Thrift is being used in a project for other Cassandra-related implementations. This means support of CQL would also be very limited with Astyanax Thrift client support. Features specific to CQL3 (e.g., collections) may not work properly with this version of Astyanax.

With Cassandra running on remote machines over multiple nodes, we can configure those remote nodes with Titan with a comma-separated list of IP addresses.

Cassandra for Backend Storage

As discussed above, Cassandra can be used as a storage backend with Titan. In this section, we will configure Titan storage options, including using Cassandra, and open a graph instance. In the following "Use Cases" section, we will demonstrate how to use Titan with Cassandra, such as with writing and reading from the graph, via some simple exercises.

1. The first thing is that we need to configure Titan for some storage options:

```
import org.apache.commons.configuration.BaseConfiguration;
import org.apache.commons.configuration.Configuration;
import com.thinkaurelius.titan.core.TitanFactory;
import com.thinkaurelius.titan.core.TitanGraph;
import com.thinkaurelius.titan.core.TitanKey;Configuration conf = new
BaseConfiguration();
conf.setProperty("storage.backend", "cassandrathrift");
conf.setProperty("storage.hostname", "localhost");
conf.setProperty("storage.port", "9160");
conf.setProperty("storage.keyspace", "twitter");
```

Table 7-1. *Titan configuration properties*

Property	Value	Description
storage.backend	cassandrathrift	Cassandra as backend storage
storage.hostname	localhost	Thrift listen_address, change according to your Cassandra server settings.
storage.port	9160	Thrift rp
storage.keyspace	Twitter	Cassandra keyspace for Titan Graph storage

Table 7-1 outlines and describes the configuration properties we used in the preceding step.

2. Next we need to open a graph instance using TitanFactory:

```
TitanGraph graph = TitanFactory.open(conf);
```

Let's further explore the Titan Java API with Cassandra via a few use cases, such as reading, writing, and batch processing data.

Use Cases

In this section, we will discuss graph traversal, reading, writing, and batch processing with graph data. Let's first discuss a scenario in which vertices and connecting edges are large in number, which is quite common with big data.

Writing Data to a Graph

After instantiating an instance of a graph, let's explore writing vertices and incident edges into a graph. We will be discussing the same Twitter example and will build a graph-based implementation for the user, its tweets, and followers.

Figure 7-19 shows a representation of a problem we will be implementing using the TitanGraph API.

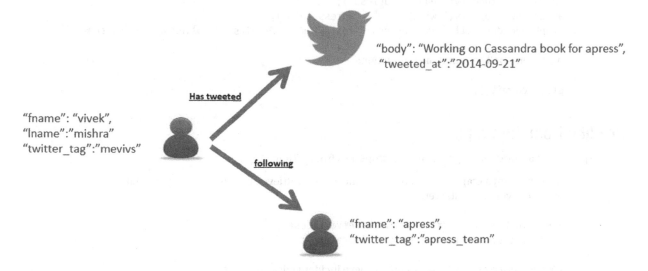

Figure 7-19. *User vivek is following apress_team and tweets about his Cassandra book on Twitter*

1. Let's add a vertex to the graph:

```
Vertex vivs = graph.addVertex(null);
vivs.setProperty("fname", "vivek");
vivs.setProperty("lname", "mishra");
vivs.setProperty("twitter_tag", "mevivs");
```

Here, Vertex is an API referred from Blueprints. The following are import statements for the preceding code snippet:

```
import com.tinkerpop.blueprints.Direction;
import com.tinkerpop.blueprints.Vertex;
```

You can assume a vertex as Java POJO and its properties as field variables.

2. Let's create another vertex for tweets and define an edge between the user and tweet vertex:

```
Vertex tweet = graph.addVertex(null);
tweet.setProperty("body", "Working on Cassandra book for apress");
tweet.setProperty("tweeted_at", "2014-09-21");
```

```
graph.addEdge(null, vivs, tweet, "has_tweeted"); // User vivs has tweets
```

3. We can also add another vertex and establish a relation of "following":

```
Vertex apress = graph.addVertex(null);
apress.setProperty("fname", "apress");
apress.setProperty("twitter_tag", "apress_team");
graph.addEdge(null, vivs, apress, "following"); // Vivs is following apress team
```

4. And then finally commit the transaction:

```
graph.commit();
```

Reading from the Graph

Let's explore a bit around reading vertices and properties from a Titan graph:

1. Reading from a graph is also fairly easy, and we can retrieve all vertices for a particular graph or even a specific vertex:

```
Iterable<Vertex> vertices = graph.getVertices();
Iterator<Vertex> iter = vertices.iterator();
```

2. We can iterate over each vertex and retrieve incident edges like this:

```
while(iter.hasNext())
{
Vertex v = iter.next();
Iterable<Edge> keys = v.getEdges(Direction.BOTH);
...
}
```

3. Each edge will have IN and OUT vertices, and we can retrieve those vertices via edges:

```
for(Edge key : keys)
{
    System.out.print(key.getVertex(Direction.IN).toString()); // will print vivs on
consle
    System.out.print("=>");
    System.out.print(key.getLabel());
    System.out.print("=>");
    System.out.println(key.getVertex(Direction.OUT).toString()); // will print tweets or
apress
}
```

The preceding reading and writing to a Titan graph provides a simple recipe for how to use Titan with Cassandra.

Cassandra is all about large data processing and analytics. It is no different when working with a graph-based model using Cassandra. So what about batch processing of data with a Titan graph database? Titan does provide support for batch data processing, and in the next example we will explore how to perform batch loading using the Titan Java API.

Batch Loading

Titan provides support for batch loading using the BatchGraph API, which can be thought of as a wrapper around TitanGraph with configurable parameters to define batch size and the type of vertex ID. We can create a BatchGraph instance as follows:

```
BatchGraph bGraph = new BatchGraph<TITAN GRAPH INSTANCE>,<VERTEX ID TYPE>,<BATCH SIZE>);
```

Using a bulk loading API, we can push a batch of records with a single database call. That way graph data loading will always be faster.

Let's explore more about bulk loading in Titan with a sample Java recipe. In this example, we will read data from a .csv file.

Figure 7-20 shows data in the format of User A following User B. Each user has fname, lname, and twitter_tag as properties of the vertex, where an edge label is **following** and contains a property value as **Cassandra.** Please note that you can find the sample .csv file with source code for this book under the src/main/resources folder.

mevivs	vivek	mishra	apress_te	apress	team	following	Cassandra
apress_team	apress	team	jhassel	Jonathan	Hassel	following	Cassandra
jhassel	Jonathan	Hassel	mevivs	vivek	mishra	following	Cassandra
jhassel	Jonathan	Hassel	apress_m	apress	marketing	following	Cassandra

Figure 7-20. *A table showing the user-to-follower relationship*

Follow these steps to complete the recipe:

1. First, the common step is to configure a graph for Cassandra:

```
import org.apache.commons.configuration.BaseConfiguration;
import org.apache.commons.configuration.Configuration;
import com.thinkaurelius.titan.core.TitanFactory;
import com.thinkaurelius.titan.core.TitanGraph;
import com.thinkaurelius.titan.core.TitanKey;Configuration conf = new
BaseConfiguration();
conf.setProperty("storage.backend", "cassandrathrift");
conf.setProperty("storage.hostname", "localhost");
conf.setProperty("storage.port", "9160");
conf.setProperty("storage.keyspace", "batchprocess");
conf.setProperty("storage.batch-loading", "true");
```

2. Let's load the sample .csv file using FileReader:

```
File file = new File("src/main/resources/bulk_load.csv");
BufferedReader reader = new BufferedReader(new FileReader(file));
```

3.   Next, create an instance of a graph and wrap it with BatchGraph:

```
TitanGraph graph = TitanFactory.open(conf);
BatchGraph bgraph = new BatchGraph(graph, VertexIDType.STRING, 1000);
```

Here 1000 is the batch size and the vertex ID is of string type.

4.   Now let's define each vertex property as a vertex key and each edge's property as a label key:

```
// prepare vertex key for each property
KeyMaker maker = graph.makeKey("twitter_tag");
    maker.dataType(String.class);
    maker.make();
    graph.makeKey("fname").dataType(String.class).make();
graph.makeKey("lname").dataType(String.class).make();

    // prepare edge properties as label
    LabelMaker labelMaker = graph.makeLabel("contentType");
    labelMaker.make();
    graph.makeLabel("following").make();
```

Here TitanKey and LabelKey are classes provided by Titan Java API, which are used to prepare vertex and edge keys.

5.   Now let's read line by line from the file and extract vertex and edge properties:

```
while (reader.ready())
    {
        String line = reader.readLine();
        StringTokenizer tokenizer = new StringTokenizer(line, ",");
        while (tokenizer.hasMoreTokens())
        {
            // System.out.println(tokenizer.nextToken());
            // twitter_tag,fname,lname,twitter_tag,fname,lname,edgeName,edgeProperty
            final String in_twitter_tag = tokenizer.nextToken();
            final String in_fname = tokenizer.nextToken();
            final String in_lname = tokenizer.nextToken();
            final String out_twitter_tag = tokenizer.nextToken();
            final String out_fname = tokenizer.nextToken();
            final String out_lname = tokenizer.nextToken();
            final String edgeName = tokenizer.nextToken();
            final String edgeProperty = tokenizer.nextToken();
        ...
    }
        }
```

6. Now create in and out vertices within an extreme out while loop (see step 5) and assign an edge as follows:

```
//in vertex
Vertex in = bgraph.addVertex(Math.random() + "");
        in.setProperty("twitter_tag", in_twitter_tag);
        in.setProperty("fname", in_fname);
        in.setProperty("lname", in_lname);

//out vertex
Vertex out = bgraph.addVertex(Math.random() + "");
        out.setProperty("twitter_tag", out_twitter_tag);
        out.setProperty("fname", out_fname);
        out.setProperty("lname", out_lname);
        //assign edge
        bgraph.addEdge(null, in, out, edgeName);
```

7. Finally we can call commit after successfully reading all records from the .csv file and populating BatchGraph:

```
bgraph.commit();
```

Here batch size is the number of vertices and edges to be loaded before we invoke the commit on the graph. One thing we should take care of is setting a moderate value as the batch size to avoid heap size issues while processing a big graph having millions or billions of edges.

The Supernode Problem

In the real world, big data-based graphs can be very large, and there can be a group of vertices having a very high number of incident edges. In graph theory, such vertices are called *supernodes*. With so many complex paths, a random traversal in a graph can lead us to such supernodes, which would badly affect the system's performance.

Figure 7-21 shows my LinkedIn social graph, where the marked vertices can be termed supernodes.

Figure 7-21. *Vivek's LinkedIn graph*

For example, if I need to traverse through all connections for a particular group (say graphDB), such traversal without indices would lead to performance issues. With incident edges indexed by label, such lookups will be much quicker.

For example, if we want to find all friends of Vivek who have joined the "graphDB" group, doing a random traversal would require searching every connection of Vivek's and then scanning groups joined by each of his friends. Imagine that Vivek has a large number of connections on LinkedIn. It can be assumed that random traversal in such a case would be a nightmare to find the desired output. But using a label-index query, it will be much quicker:

```
g.query ().has("friends of",EQUAL, "vivek")..has("group",EQUAL,"graphDB").vertices();
```

This query is a label-indexed query, which searches all of Vivek's friends using "friends of" and then searches the remaining subset for graphDB using the "group" indexed edge.

We will explore this further in next section of this chapter.

Faster Deep Traversal

Deep traversal means going to the n'th level in the hierarchy of a graph. Let's take the example we saw in the preceding section of my LinkedIn social graph, where I can query my **connections** for common group interests. Depending on the data volume and number of incident edges, iterating over each vertex via vertex query is probably not a good idea. Titan provides support for multiple-vertex queries, where multiple-vertex queries can be combined and send a single combined query to a graph database. That way retrieval of data will be a lot faster because we will be hitting the server only one time.

Let's explore how to achieve faster deep traversal using a multiple-vertex query in Titan with a Hindu mythological epic called Ramayana. In this example we will try to find **son of** relationships at the leaf level. Figure 7-22 shows the family tree of Rama and their ancestors.

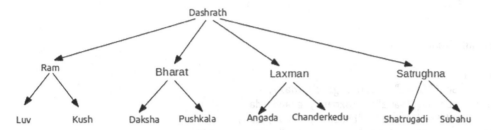

Figure 7-22. *The Ramayan family graph*

One way to find **son of** relationships at each level is to iterate through each level like this:

```
private static void iterateToLeaf(Vertex dasratha)
{

    System.out.println("Finding sons for::" + dasratha.getProperty("fname"));
    Iterable<Vertex> immediateSons = dasratha.getVertices(Direction.IN, "son of");
    Iterator<Vertex> iter = immediateSons.iterator();
    // one way is
    while (iter.hasNext())
    {
        Vertex v = iter.next();
        // recursive call
        iterateToLeaf(v);

    }
}
```

In the preceding code snippet, we need to invoke with the root vertex object, i.e., `dasaratha`, and then the recursive call will iterate through each vertex on each level. For smaller graphs it may work, but for large data and big data graphs, it is not a feasible solution. This is where multiple-vertex query comes in very handy and performant.

Let's walk through a few code snippets to execute a multivertex query using Titan. You can find a complete GraphTraversalRunner.java example with the attached source code.

1. First let's add dasaratha as root vertex:

```
Vertex das = add("fname", "dasaratha", graph);

private static Vertex add(final String propertyName, final String value, TitanGraph graph)
{
    Vertex vertex = graph.addVertex(null);
    vertex.setProperty(propertyName, value);

    return vertex;
}
```

2. Next, add dasaratha's son:

```
// add dasratha's son.
Vertex ram = addSon("fname", "ram", graph, das);
Vertex laxman = addSon("fname", "laxman", graph, das);
Vertex bharat = addSon("fname", "bharat", graph, das);
Vertex shatrugna = addSon("fname", "shatrugna", graph, das);
    private static Vertex addSon(String propertyName, String value, TitanGraph graph, Vertex
      father)
    {
        Vertex son = add(propertyName, value, graph);
        graph.addEdge(null, son, father, "son of");
        return son;
    }
```

3. Repeat step 2 for ram, laxman, bharat, and shatrugna:

```
// ram's son
addSon("fname", "luv", graph, ram);
addSon("fname", "kush", graph, ram);

// bharat's son
addSon("fname", "Daksha", graph, bharat);
addSon("fname", "Pushkala", graph, bharat);

// laxman's son
addSon("fname", "Angada", graph, laxman);
addSon("fname", "Chanderkedu", graph, laxman);

// Shatrugna's son
addSon("fname", "Shatrugadi", graph, shatrugna);
addSon("fname", "Subahu", graph, shatrugna);
```

4. Finally store the complete hierarchy:

```
graph.commit();
```

5. Now to fetch all vertices with `Direction.IN` (means incoming edges) and having the label son of using multivertex support, we need to execute the following query:

```
//prepare multi vertex query
TitanMultiVertexQuery mq = graph.multiQuery();
mq.direction(Direction.IN).labels("son of");

mq.addVertex((TitanVertex) das);// add root
mq.addVertex((TitanVertex) ram);
mq.addVertex((TitanVertex) bharat);
mq.addVertex((TitanVertex) laxman);
mq.addVertex((TitanVertex) shatrugna);

//execute multi vertex query
Map<TitanVertex, Iterable<TitanVertex>> dfsResult = mq.vertices();

//iterate through result and print
for (TitanVertex key : dfsResult.keySet())
{
    System.out.println("Finding son of" + key.getProperty("fname"));
    Iterable<TitanVertex> sons = dfsResult.get(key);
    Iterator<TitanVertex> sonIter = sons.iterator();
    while (sonIter.hasNext())
    {
        System.out.println(sonIter.next().getProperty("fname"));
    }
}
```

This way, we can perform faster deep traversal with Titan.

We have covered most of the important features supported by Titan graph database. Sample code snippets shared in this chapter should enable readers to use Titan with Cassandra, such as building social graphs, network graphs, or running graphs like queries. As far as the future of graph databases is concerned, the next chapter will discuss a report published about active development which is happening in the graph database world.

With this, we conclude our discussion around graph databases and how Titan can be integrated with Cassandra. For more details and supported feature sets you may refer to https://github.com/thinkaurelius/titan/wiki.

Summary

To summarize, topics covered in this chapter include:

- Introduction to graphs
- Understanding TinkerPop and Blueprints
- Titan database ecosystem
- Titan with Cassandra

The next chapter will walk you through the performance tuning and compaction techniques available with Cassandra.

CHAPTER 8

■ ■ ■

Cassandra Performance Tuning

As storage hardware has become cheaper and more efficient in terms of capacity, organizations are able to choose less expensive, lowmaintenance storage options for their terabytes of data. Obviously cost effectiveness has always been an important factor behind technology selection. Storage and retrieval of large data are requirements, but an economical solution should not be at the cost of performance. Data, whether large or small, isn't good unless it can be used effectively and have analytics applied to it.

With proven opportunities from real-world big data-related use cases, many organizations and startups are exploring business ideas around large data. Real-time feeds are very much in demand for companies targeting an online audience. Applications nowadays deal with gigabytes of data per second to be processed and analyzed, which clearly shows that performance is an important component for modern big data-based applications.

In this chapter we will discuss

- Key performance indicators
- Cassandra cache configurations
- Discuss Bloom filters and garbage collection
- Cassandra stress testing
- Yahoo Cloud Serving Benchmarking

Understanding the Key Performance Indicators

Several key performance indicators (KPIs) will be discussed in this section, including:

- CPU and memory consumption
- Heavy read/write throughput and latency
- Logical and physical reads

Inconsistent performance or performance degradation are main points where applications need performance tuning. This is where front-end applications and back-end databases need to be tuned from a performance perspective.

To begin, let's take an initial look at several KPIs that inform how we'll configure Cassandra to enhance performance.

CPU and Memory Utilization

CPU and memory utilization are very important from a performance perspective. There are tools like JConsole, JProfile, and YourKit to monitor application's performance via monitoring CPU and memory footprints. It's worth mentioning that many Cassandra server processes take off-heap memory; hence, while tuning heap configuration for applications running locally on the same box, we need to make sure that memory shortage or leakage does not occur and bring instability to the application. A low heap size configuration would trigger frequent garbage collection activities. Avoiding the garbage collection cycle is definitely not possible but we can tune it for better response time. For example, keeping the younger generation big would accommodate more live objects and would lead to long GC (garbage collection) cycles. In general garbage activity over younger generation objects will be fast and overhead on response time would be almost negligible. But Full GC is sort of a "stop the world" event and can affect an application's performance badly. We need to make sure to allocate proper memory to Eden Spaces and later temporary and permanent spaces in the memory pool. It is recommended to keep smaller permanent generation space.

Heavy Read/Write Throughput and Latency

Discussion of big data processing is all about "walking the talk." Application must be enabled to perform faster ingestion and reads irrespective of data volume. Read/write latency is another important aspect of performance monitoring and tuning. Modern applications demand real-time or near real-time data processing with minimal latency. Hence while adopting technology and databases we must keep this in mind to avoid such performance overheads.

Logical and Physical Reads

Logical reads mean reading rows from the data cache, whereas physical reads seek disc access. The higher side on logical reads tends to have a big memory footprint, and, similarly, frequent physical reads introduce read latency and can introduce performance overhead. Hence it is important to strike a balance between both. Cassandra provides row and key cache along with memtables to reduce physical reads. We will discuss these in detail in this chapter.

Next, let's discuss various Cassandra-specific configurations. During the latter part of this chapter we will cover performance monitoring and tuning for front-end applications, as well.

Cassandra Configuration

Cassandra offers multiple configurations that can be tweaked for heavy read/write load. Let's explore them one by one, starting with cache configuration.

Data Caches

Data caches keep frequently accessed rows in memory to avoid disc access at the server side. In n-tier architecture, data caching can also be implemented at the client side (e.g., Ehcache). The concept of data caching is not new and has been there with traditional RDBMS, as well. Data cache is very useful with read operations, and it is recommended to opt for database level caches for frequently fetched data. Cassandra offers both key and row caches. Figure 8-1 shows a flow chart diagram of how these caches work while reading data from the Cassandra database.

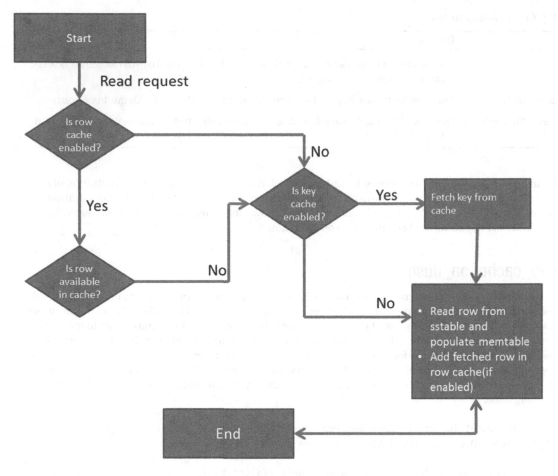

Figure 8-1. *How a cache works with the Cassandra server upon receiving a read request*

Row and key caches are helpful in retrieving rows most frequently used. Keeping a moderate size of row caches and a higher number of key caches can result in better performance and less latency. Let's discuss these cache configurations with Cassandra.

Cache Directory

The default cache location is /var/lib/cassandra/saved_caches. We can also modify it in cassandra.yaml:

saved_caches_directory: /var/lib/cassandra/saved_caches

Key Cache

With key cache, references to actual row keys will be cached in memory for each column family. Having a large key cache is recommended for heavily used, read-based applications, obviously within permissible JVM settings. Various key cache configurations that can be configured with cassandra.yaml are described in Table 8-1.

Table 8-1. *Key Cache Configurations*

Name	Description
key_cache_size_in_mb	Key cache size in megabytes. By default it is 5% of heap or 100MB (whatever is less). It can be disabled by setting value as 0.
key_cache_save_period	Time period to save key cache on configured cache directory. Default is 4 hours
key_cache_keys_to_save	Number of keys to be saved, by default it is disabled but recommended setting is to set it for high value.

With 2.0 onwards, key cache has been moved to off-heap, which reduces the time spent on serialization of keys significantly. We will discuss off-heap versus on-heap later in this chapter. As discussed earlier, memory utilization is one of the important KPIs for performance and while configuring row caches, we must avoid Full GC activities, regardless of whether the key cache has been moved to off-heap now.

populate_io_cache_on_flush

By default it is disabled and set to false. Prior to Cassandra 1.2.x, it was supported at node level but now it is possible to enable it at column family level. The purpose of this flag is to enable/disable page cache at the column family during memtable flush and compaction. If set to true, it will keep a cache of the specific column family's data in memory. As discussed in Chapter 1, memtables keep flushing data onto the disc in the form of sstables. If this parameter is set to true Cassandra keeps a page cache for all rows of this column family. Please note that page cache refresh only happens in case of flushing and compaction to keep the updated data on the node for the respective column family. Users need to make sure that data size fits well within the node's memory. We can enable it while creating the table or altering the existing table:

```
create table tweets(tweet_id text primary key,body text) with caching='rows_only' and
bloom_filter_fp_chance=0.004 and populate_io_cache_on_flush='true';  // create table

alter table user with populate_io_cache_on_flush='true'; //alter table
```

It is recommended to keep this true for frequently queried column families which can fit well within memory for better read throughput. Figure 8-2 shows the table the user created with this attribute set to true (see marked line).

```
cqlsh:twitter> describe table user;

CREATE TABLE user (
  user_id text,
  fname text,
  PRIMARY KEY (user_id)
) WITH
  bloom_filter_fp_chance=0.004000 AND
  caching='KEYS_ONLY' AND
  comment='' AND
  dclocal_read_repair_chance=0.000000 AND
  gc_grace_seconds=864000 AND
  index_interval=128 AND
  read_repair_chance=0.100000 AND
  replicate_on_write='true' AND
  populate_io_cache_on_flush='true' AND
  default_time_to_live=0 AND
  speculative_retry='99.0PERCENTILE' AND
  memtable_flush_period_in_ms=0 AND
  compaction={'class': 'SizeTieredCompactionStrategy'} AND
  compression={'sstable_compression': 'LZ4Compressor'};
```

Figure 8-2. *User table enabled with populate cache on I/O operations set to true*

Row Cache

With row level cache enabled, the entire row is available in cache. Row cache requires a lot of space in comparison to key level cache. With Cassandra 1.2 onwards row cache can be moved to off-heap, but deserialization of rows still happens temporarily in-heap. Table 8-2 shows several row cache configurations and their descriptions.

Table 8-2. *Row Cache Configurations*

Name	Description
row_cache_size_in_mb	By default it is disabled and set to 0.
row_cache_save_period	By default row cache is disabled, enabling it would improve on cold start and cache can be expensive and memory intensive.
row_cache_keys_to_save	By default disable. Number of keys of row cache to be saved.

Row cache comes in very handy and reduces disk read significantly for frequently accessed rows. But it consumes a lot of space as well. It is recommended to enable row cache if the application's requirement is to fetch a complete row instead of select columns very often.

To enable a cache for a specific column family, we can enable it while creating a column family as follows:

```
create table user(user_id text primary key,fname text) with caching='keys_only';
create table tweets(tweet_id text primary key,body text) with caching='rows_only';
```

In the preceding script for a column family, user caching has been enabled for keys but for the tweets table it is enabled for rows. It is recommended to enable row cache where the application would require reading entire rows rather than selected columns. Obviously with the selected column approach, having row cache enabled would have an adverse impact and be a waste of in-memory space by keeping rows within memory for no purpose.

The following code snippet shows how to configure these cache parameters in the `Cassandra.yaml` file:

```
# 50 MB keys to be saved in cache.
key_cache_size_in_mb: 50
# 5 minutes or 3600 seconds to keep keys cached in cache
key_cache_save_period: 3600
# 100 keys to save in key cache
key_cache_keys_to_save: 100
#keep 100 mb of rows in cache
row_cache_size_in_mb: 100

# 5 minute or 3600 seconds, duration to keep rows in cache.
row_cache_save_period: 3600
#100 number of keys from row cache to be saved.
row_cache_keys_to_save: 100
#saved cache directory.
saved_caches_directory: /var/lib/cassandra/saved_caches
```

Bloom Filters

A Bloom filter is a data structure to manage whether an element is present in a set or not. It was conceived by Burton Bloom in 1970 to find out the probability of whether an element exists in a set or not. With a Bloom filter enabled it will return whether an element is "definitely not in set" or "may exist in a set." A false positive (FP) means the element may exist in a set when it doesn't, and a false negative means an element definitely is not present in a set when it is.

Each table in Cassandra contains a Bloom filter. With a false positive chance ratio value, it checks for columns in a row within the sstable that may exist but for which a false negative is definitely not possible. Here false negative means that columns of the row exist but the Bloom filter returns negative.

Setting the Bloom filter FP ratio higher would mean less memory consumption and ensure that false negatives would never occur (e.g., No disk i/o for non-existing keys) Range of Bloom filter FP ratio is .000744 to 1.0. The result of setting a false positive ratio chance to a higher level is that there is the possibility finding a column in the sstable, but there are no disk reads for negative scenarios.

You can set Bloom filter while creating column family or can also update the column family like this:

```
create table tweets(tweet_id text primary key,body text) with caching='rows_only' and
bloom_filter_fp_chance=0.004;

alter table user with bloom_filter_fp_chance=0.004;
```

Off-Heap vs. On-Heap

Heap offloading or off-heap is directly allocating memory from the operating system, whereas on-heap memory objects are managed by the JVM itself. Cassandra also loads row and key caches in off-heap memory, which means these objects are serialized on a non-Java heap. That way deserialization will be much faster.

It is recommended to keep heap size moderate, as heap takes memory only from system memory. It has been observed that it is better to not keep it more than 6 GB (if system memory is more than 6 GB). You can fetch maxMemory by using Java runtime:

```
Runtime rTime = Runtime.getRuntime();
 long maxSize = rTime.maxMemory();
```

If system memory is less than 2 GB, it is recommended to keep 50% of it.

Cassandra requires that the Java Native Access (JNA) library be installed for heap offloading caches.

To enable the JNA library, you need to download the JNA.jar from https://github.com/twall/jna and set on the classpath. That's it!

We can also set off-heap memory allocator within cassandra.yaml:

```
memory_allocator: NativeAllocator
possible values are NativeAllocator, JEMallocAllocator.
```

To install JEMallocAllocator, you need to download and install the jemalloc library. Source for jemalloc is available over GitHub (https://github.com/jemalloc/jemalloc).

Installing and Configuring jemalloc

The steps to install jemalloc are

1. First clone and download the source from github:

    ```
    git clone https://github.com/jemalloc/jemalloc.git
    ```

2. Change the directory to jemalloc and run configure:

    ```
    cd jemalloc
    ./configure
    ```

3. Next, install it by running the following:

    ```
    make
    make install
    ```

4. After successfully installing jemalloc, we need to configure installed libraries for Cassandra in cassandra-env.sh:

    ```
    export JEMALLOC_HOME= jemalloc/
    export LD_LIBRARY_PATH=<JEMALLOC_HOME>/lib/
    JVM_OPTS="$JVM_OPTS -Djava.library.path=<JEMALLOC_HOME>/lib/"
    ```

Garbage Collection

Cassandra does explicit garbage collection, and server logs can be monitored for the GCInspector class for more information. You may opt to disable explicit garbage collection using XX:+DisableExplicitGC, but it is advisable to do it only for experimental purposes. Alternatively configuring gcInterval settings are also an option (sun.rmi.dgc.server.gcInterval and sun.rmi.dgc.client.gcInterval). It represents the maximum interval (in ms) for Java runtime RMI (Remote Method Invocation) to call for garbage collection of unused and soft references. By default it is set to 1 hour.

We can tune garbage collection settings in `cassandra-env.sh` available in the `$CASSANDRA_HOME/conf` folder. A few of the options are

```
# GC logging options -- uncomment to enable
 JVM_OPTS="$JVM_OPTS -XX:+PrintGCDetails"
 JVM_OPTS="$JVM_OPTS -XX:+PrintGCDateStamps"
 JVM_OPTS="$JVM_OPTS -XX:+PrintHeapAtGC"
 JVM_OPTS="$JVM_OPTS -XX:+PrintTenuringDistribution"
 JVM_OPTS="$JVM_OPTS -XX:+PrintGCApplicationStoppedTime"
 JVM_OPTS="$JVM_OPTS -XX:+PrintPromotionFailure"
 JVM_OPTS="$JVM_OPTS -XX:PrintFLSStatistics=1"
 JVM_OPTS="$JVM_OPTS -Xloggc:/var/log/cassandra/gc-`date +%s`.log"
# If you are using JDK 6u34 7u2 or later you can enable GC log rotation
# don't stick the date in the log name if rotation is on.
 JVM_OPTS="$JVM_OPTS -Xloggc:/var/log/cassandra/gc.log"
 JVM_OPTS="$JVM_OPTS -XX:+UseGCLogFileRotation"
 JVM_OPTS="$JVM_OPTS -XX:NumberOfGCLogFiles=10"
 JVM_OPTS="$JVM_OPTS -XX:GCLogFileSize=10M"
```

Hinted Handoff

This feature comes in very handy when one or more replica nodes are down or not available while writing a row. In such a scenario, the coordinator node would keep a copy of data in the system.hints table until the replica node(s) are up and running. Generally it is a nice feature, but if such failure happens on multiple nodes, it would increase heap size on the coordinator node. The system.hints table holds the IP addresses of the replica node and the actual data that needs to replay when the replica node comes alive. During the bootstrap process, system tables would be loaded in memory so a larger number of failures would mean bigger data in memory on the coordinator node. Such scenarios may also result in data loss (until repair). By default this feature is enabled. Disabling it might result in relatively smaller heap sizes and less GC activities. But for durable writes it is advisable to keep this enabled. By default the coordinator will keep data in the system.hints table for 3 hours, i.e., a default value of `max_hint_window_in_ms` configuration in the Cassandra.yaml file. If the replica node doesn't come up in this configured window time, it would be assumed as a dead node and later when it comes up it will need to run repair and replicate data from other nodes.

Heap Size Configuration

Generally for applications, it is recommended to keep heap size in the range of 4–8 GB; but if memory is less than 4 GB, we still should use half of it (~2 GB) for I/O operations. You can modify JVM settings for heap size configuration in `cassandra-env.sh` (look for `JVM_OPTS`).

```
JVM_OPTS="$JVM_OPTS $JVM_EXTRA_OPTS" -Xms1G  -Xmx1G  -XX:+HeapDumpOnOutOfMemoryError^
-XX:+UseParNewGC  -XX:+UseConcMarkSweepGC
```

Cassandra Stress Testing

Cassandra stress is a Java-based tool to test server performance under heavy read/write load. With binary distribution it is available under the $CASSANDRA_HOME/tools/bin directory. The most commonly used options for stress testing are

- Write only
- Read only
- Search range slice

The stress tool is meant to monitor a database's performance under heavy reads, heavy writes, and reads via nonprimary key columns. With various available parameters the user can experiment and analyze Cassandra's capability as per their requirement using this tool to search all available options with this utility you can use the help option:

```
$CASSANDRA_HOME/tools/bin/cassandra-stress -h
```

Write Mode

Let's start with the default mode, i.e., write mode, where we will perform stress testing of the Cassandra server using this tool. With this mode we will be creating a column family with a predefined column size and total number of write operations.

Let's start with a simple write-only mode with a simple command:

```
$CASSANDRA_HOME/tools/bin/cassandra-stress
```

The output of this command would be similar to Figure 8-3.

```
vivek@vivek-Vostro-3560:~/software/apache-cassandra-2.0.4/bin$ ../tools/bin/cassandra-stress
Created keyspaces. Sleeping 1s for propagation.
total,interval_op_rate,interval_key_rate,latency,95th,99.9th,elapsed_time
170925,17092,17092,1.1,4.5,287.6,10
385212,21428,21428,1.3,4.3,282.4,20
591433,20622,20622,1.3,4.5,282.8,30
803491,21205,21205,1.4,4.9,282.8,40
995098,19160,19160,1.4,5.2,284.5,50
1000000,490,490,1.4,5.2,284.5,50


Averages from the middle 80% of values:
interval_op_rate         : 20086
interval_key_rate        : 20086
latency median           : 1.3
latency 95th percentile  : 4.6
latency 99.9th percentile : 283.9
Total operation time     : 00:00:50
END
```

Figure 8-3. *Output from running the stress tool in default write mode*

A few key points to monitor from the output shown in Figure 8-3 include

- **total**: the total number of operations.

- **interval_op_rate**: the number of operations performed during that interval.

- **interval_key_rate**: the number of rows written during the interval.

- **latency**: the average write latency for each operation in that interval.

- **elapsed_time**: total number of seconds spent during run.

Here the default host and port are localhost and 9160, respectively. By default, the number of records to be inserted is one million.

You can also overwrite default host, port, and number of records as follows:

```
$CASSANDRA_HOME/tools/bin/cassandra-stress -n <NO_OF_RECORD> -d <HOST> -p <PORT>
```

Total time taken to run this stress test is ~50 seconds.

Let's quickly connect to CQL3 to verify stored rows. The stress test will create a default keyspace called **"Keyspace1"** and a table called **data**. To get a count on the total number of inserted rows we can run the following command:

```
use "Keyspace1" ;
select count(*) from "Standard1" limit 10000000;
```

Figure 8-4 shows the result.

```
cqlsh:Keyspace1> select count(*) from "Standard1" limit 10000000;

 count
---------
 1000000

(1 rows)
```

Figure 8-4. *A cql query to find the total number of records in the Standard1 column family created via the stress tool*

Reader can use the describe table command to describe the default columns and their data types. Here the default column size is 5 and all are of type blob.

Similar to other default values, it is also possible to overwrite column count, size, family type, thread count, and random row key generation. Let's examine the sample output by running insert only for 20 and 30 threads, respectively.

With 20 threads

```
$CASSANDRA_HOME/tools/bin/cassandra-stress -d localhost -p 9160 -o insert -t 20 -r -S 50 -U UTF8Type
-y Standard
Averages from the middle 80% of values:
interval_op_rate          : 16657
interval_key_rate         : 16657
latency median            : 2.8
latency 95th percentile   : 11.5
latency 99.9th percentile : 646.7
Total operation time      : 00:01:01
END
```

Figure 8-5 shows a JConsole overview chart of CPU usage, heap, and other resource utilization with 20 threads.

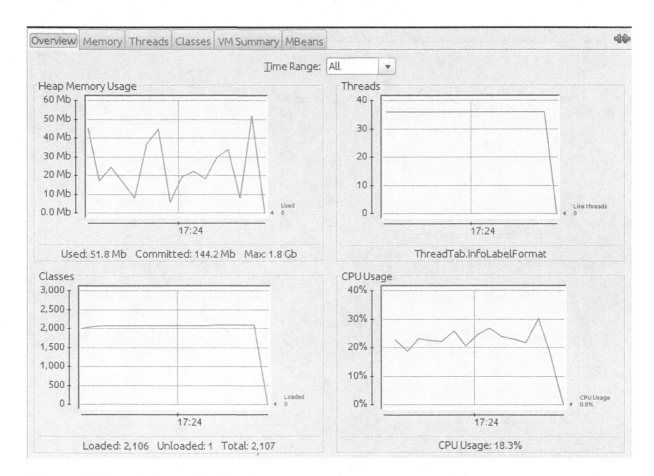

Figure 8-5. *A JConsole chart with 20 threads*

With 30 threads

```
$CASSANDRA_HOME/tools/bin/cassandra-stress -d localhost -p 9160 -o insert -t 30 -r -S 50 -U UTF8Type
-y Standard
```

```
Averages from the middle 80% of values:
interval_op_rate          : 18467
interval_key_rate         : 18467
latency median            : 0.5
latency 95th percentile   : 2.8
latency 99.9th percentile : 122.8
Total operation time      : 00:01:04
```

Figure 8-6 shows a JConsole overview chart of CPU usage, heap, and other resource utilization with 30 threads.

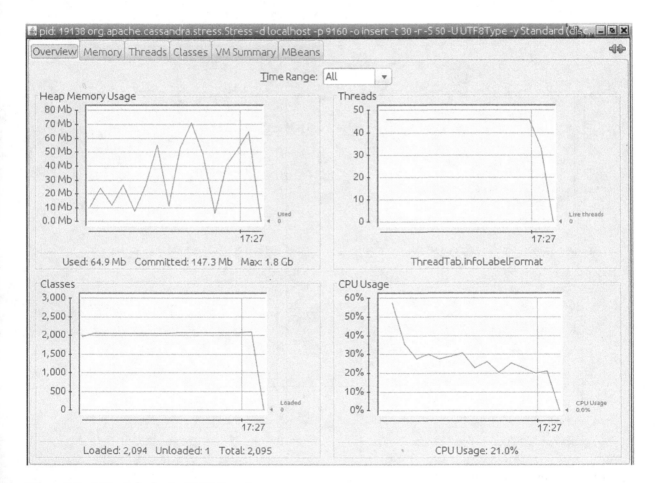

Figure 8-6. *A JConsole chart with 30 threads*

With a different number of threads it is clear that parameters can vary in terms of resource utilization and performance. It is advisable to use stress tests to monitor the system's limit based on the application's SLA.

Read Mode

With read mode, let's test the Cassandra server for with a heavy read-only operation. Here we will perform a stress test over the Cassandra server to monitor its performance with various parameters.

The simplest command to start a read stress test is

```
$CASSANDRA_HOME/tools/bin/cassandra-stress -o read
```

Figure 8-7 shows the result.

```
vivek@vivek-Vostro-3560:~/software/apache-cassandra-2.0.4/bin$ ../tools/bin/cassandra-stress -o read
total,interval_op_rate,interval_key_rate,latency,95th,99.9th,elapsed_time
60178,6017,6017,4.4,29.4,111.6,10
148215,8803,8803,3.5,26.6,117.7,20
259379,11116,11116,3.0,22.1,117.7,30
425796,16641,16641,2.4,13.9,117.7,40
681714,25591,25591,1.8,9.1,110.2,50
973080,29136,29136,1.6,7.0,63.4,60
1000000,2692,2692,1.5,6.2,63.4,61


Averages from the middle 80% of values:
interval_op_rate          : 15537
interval_key_rate         : 15537
latency median            : 2.7
latency 95th percentile   : 17.9
latency 99.9th percentile : 115.9
Total operation time      : 00:01:01
END
```

Figure 8-7. *Output from running the stress tool in a heavy read operation*

The total time taken to run the read stress test is slightly higher than the write stress test. Since we have tried tweaking the stress test for other parameters, such as column size, number of columns, and number of threads in write mode, here we'll leave it for readers to try the available options with read mode as an exercise.

Monitoring

For monitoring the progress of stress testing, you may use jmxterm or JConsole for monitoring memory usage and threads. Figure 8-8 shows one such image in JConsole captured during the previous run of read mode.

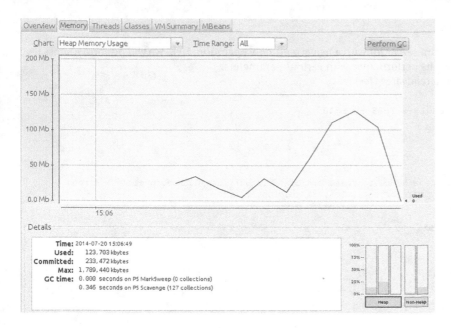

Figure 8-8. *An image in JConsole captured during the run of read mode*

For more details on how to configure and connect JConsole with Cassandra you may refer to Chapter 1.

For third-party tools, DataStax's OpsCenter provides detailed information about running the Cassandra cluster. For more details on usage of the DataStax OpsCenter please refer to www.datastax.com/what-we-offer/products-services/datastax-opscenter.

Compaction Strategy

During Cassandra stress testing, we have seen that writes in Cassandra are blazingly fast. The reasons behind this are that there are no updates and all columns are directly flushed into sstables. This means that at some point there would be many versions of column values which may be present in form of multiple sstables. Similarly physical deletes would not happen in real time but those are logically deleted instantly. Such rows are called *tombstone* or *obsolete rows*. The process to remove such columns and free some space at the server side is known as *compaction strategy*.

Compaction always runs in the background with async mode. Cassandra offers two types of compaction strategies:

- Size-tiered compaction strategy

- Leveled compaction strategy

Size-Tiered Compaction Strategy (STCS)

Size-tiered compaction strategy (STCS) is the default compaction strategy. Based on configuring min_compaction_threshold (by default 4), sstables of similar size would be merged and compacted in a single table. Clearly for heavy writes with not many reads/updates, this is the preferred strategy. Although all the data for one particular row key will reside on the same data node, it still may be scattered across multiple sstables. With such a compaction strategy it is highly possible that redundant compaction for the same columns may happen until the process is complete. With frequent updates for the same row key, STCS could be problematic as a compaction strategy.

Leveled Compaction Strategy (LCS)

Leveled compaction strategy's (LCS) implementation is heavily inspired by Google's LevelDB (http://code.google.com/p/leveldb). The purpose of LCS is to avoid redundant compaction for columns of a row key. With LCS, each level of sstables would never overlap and guarantees data loss on read. With every level up, the compaction process will take place and keep adding the sstables from previous ones. Here it would require enough space to keep the largest level within memory. LCS performs more I/Os than STCS.

You can define a leveled compaction strategy while creating a table like this:

```
create table tweets(tweet_id text primary key,body text) with caching='rows_only' and compaction
='LeveledCompactionStrategy' and populate_io_cache_on_flush='true';  // create table
```

LCS is recommended if the read proportion is somewhat similar to higher write proportions or frequent updates are expected.

Yahoo Cloud Serving Benchmarking

The Yahoo Cloud Serving Benchmark (YCSB) project is the industry standard for evaluation of various cloud data stores and key-value data stores using various workloads in parallel execution. It was released by Yahoo's research labs in 2010 and is hosted at https://github.com/brianfrankcooper/YCSB.

YCSB is totally decoupled and extensible to add any new datastore-specific client.

Configuring and running YCSB is fairly easy. In this recipe we will discuss configuring the YCSB framework with Cassandra.

1. First, you can build the jar from the source itself or download the tarball from

    ```
    https://github.com/downloads/brianfrankcooper/YCSB/ycsb-0.1.4.tar.gz
    ```

2. Extract the tarball and change the directory:

    ```
    cd ycsb-0.1.4
    ```

3. By default YCSB would insert data into the **usertable** keyspace and in the table named **data**. So run the following command from Cassandra-cli to create data definitions:

    ```
    create keyspace usertable;
    use usertable;
    create column family data;
    ```

4. Next create a property file (e.g., Cassandra-ycsb.properties) with the following properties:

    ```
    hosts=localhost
    port=9160
    fieldcount=4
    threads=9
    recordcount=1000
    operationcount=1000
    ```

```
workload=com.yahoo.ycsb.workloads.CoreWorkload
readallfields=true
readproportion=0.5
updateproportion=0.5
scanproportion=0
insertproportion=0
```

Here, hosts and port are Cassandra Thrift host and rpc_port settings. fieldcount is the total number of fields to be stored per row. threads determine the number of parallel executions. Also readproportion and updateproportion can be tweaked during the tests. For example for heavy write only scenarios, you may keep insertproportion to 1.0 and rest as 0.0! Similarly for heavy read scenarios can set readproportion to 1.0.

5. Finally, run this command:

```
bin/ycsb load cassandra-10 -P /home/vivek/source/book_source/ycsb/casssandra-ycsb.
properties -threads 10
```

The output from running the preceding command will be a fairly large. Figure 8-9 shows what is interesting from a performance perspective.

```
SLF4J: Failed to load class "org.slf4j.impl.StaticLoggerBinder".
SLF4J: Defaulting to no-operation (NOP) logger implementation
SLF4J: See http://www.slf4j.org/codes.html#StaticLoggerBinder for further details.
[OVERALL], RunTime(ms), 233.0
[OVERALL], Throughput(ops/sec), 4291.845493562232
[INSERT], Operations, 1000
[INSERT], AverageLatency(us), 1545.354
[INSERT], MinLatency(us), 236
[INSERT], MaxLatency(us), 23316
[INSERT], 95thPercentileLatency(ms), 3
[INSERT], 99thPercentileLatency(ms), 10
[INSERT], Return=0, 1000
```

Figure 8-9. Output of running load using the YCSB Cassandra project

In Figure 8-9, the key points are average throughput and latency. For this run there are 4291 operations per second and latency is 1545 us for a total of 1000 operations.

YCSB offers a great way to evaluate databases from the client perspective. You may do a quick analysis of multiple clients' performance against Cassandra. Or you can also use it for performance comparison among multiple datastores. The framework is completely decoupled which makes it fairly easy to add clients for new datastores or add multiple clients for same datastore.

Coming back to read operation analysis via YCSB, we can run it as

```
bin/ycsb run cassandra-10 -P /home/vivek/source/book_source/ycsb/casssandra-ycsb.properties -threads 10
```

With YCSB there are two modes of running load and run. With load it will store data into backend and throughput and latency would be measured. Whereas with run mode we can tweak it for heavy read or read/update proportions (see `Cassandra-ycsb.properties`).

YCSB comes with two types of distribution: zipfian and uniform. Continuous uniform distribution would refer to symmetric distribution of data for each column at each interval. Whereas zipfian distribution is based on power law distribution where load and throughput are inversely proportional to each other. For more details on zipfian distribution please refer to `http://en.wikipedia.org/wiki/Zipf's_law`.

Summary

As we reach the end of this chapter, here are a few key performance checks to make if an application's throughput is consistently degraded over time:

- Monitor GC and examine the Cassandra server logs for the `GCInspector` class.

- If data nodes are running out of space, add new nodes. We will discuss this further in the next chapter.

- Monitor `bloom_filter_fp_chance` and cache configuration.

In the next chapter, we will cover more about Cassandra administration and monitoring.

CHAPTER 9

■ ■ ■

Cassandra: Administration and Monitoring

Database administration means managing a set of activities like ensuring database availability, managing security, tuning, and backups. The details of "administration," however, have undergone significant change.

Prior to NoSQL world evolution, database administrators (DBAs) were considered to be one of the most critical members in the team. But with the rise in NoSQL's popularity, administrative effort and the need for a dedicated team of database administrators have been reduced significantly. In Chapter 1 we discussed changing the gears from RDBMS to NoSQL world. Let's revisit a few of those key differences:

- No joins

- No relational schema

- No static schema

That means administration efforts, at least from an application design perspective, definitely are reduced, and the "Who crashed the server" fight among developers and DBAs may soon cease to exist.

But that doesn't mean database administration is not required at all. NoSQL means "*Not only* SQL" and not "*No* SQL." We may be required to spend less effort on application modeling, but core points like security, tuning, and so forth, still need to be addressed. With traditional RDBMS, we might embed application logic at the database level, whereas with NoSQL this currently is often not required or is at least less often necessary. However, as the NoSQL world is evolving and maturing, more administrative effort will be necessary to manage things that are more important and involved than adding indexes and creating tables.

In this chapter, the following topics will be covered:

- Adding new nodes to the Cassandra cluster

- Replacing a node

- Database backup and restoration

- Monitoring tools

DBA SKILLS

As the database world changes, understanding the requisite skillsets for administrators is important. For proper and effective database administration, the following basic skills are required:

- The ability to design and model databases based on a provided logical blueprint.

- BA proficiency in managing database backup and restoration procedures.

- The ability to manage data integrity and its high availability.

- The skill to ensure data security. Database administrators are responsible for managing data security layers among various environments, such as testing, staging, and production environments.

Adding Nodes to Cassandra Cluster

In Chapter 1, we discussed configuring a three-node Cassandra cluster on a single machine. In this section we will first start the Cassandra cluster with two nodes and later will add another node.

1. First, for Random partitioner let's generate tokens for all three nodes using the token-generator tool. This tool can be found under $CASSANDRA_HOME/tools/bin folder.

```
vivek@vivek-Vostro-3560:~/software/local-cluster/node1$ tools/bin/token-generator

RandomPartitioner
Token Generator Interactive Mode
--------------------------------

 How many datacenters will participate in this Cassandra cluster? 1
 How many nodes are in datacenter #1? 3

DC #1:
  Node #1:                                 0
  Node #2:    56713727820156410577229101238628035242
  Node #3:   113427455640312821154458202477256070484
```

Here software/local-cluster/node1 stands for $CASSANDRA_HOME. If you choose to use Murmur3Partitioner, you can generate token values by executing the following command:

```
python -c 'print [str(((2**64 / 3) * i) - 2**63) for i in range(3)]'
```

It would give this output:

```
['-9223372036854775808', '-3074457345618258603', '3074457345618258602']
```

The next step is to configure three nodes for cluster setup. For detailed instruction about how to set this up, please refer to the "Configuring Multiple Nodes over a Single Machine" section in Chapter 1. In brief, we need to modify configurations such as listen_address, rpc_port, jmx_port, the data directory, the commit-log directory, and native_protocol in corresponding Cassandra.yaml.

2. Let's start both the nodes and check the ring status:

```
vivek@vivek-Vostro-3560:~/software/local-cluster/node2$ bin/nodetool -h 127.0.0.2 -p 7200 ring
```

```
Datacenter: datacenter1
==========
Address     Rack       Status State   Load       Owns       Token
                                                             -3074457345618258603
127.0.0.1   rack1      Up     Normal  36.06 KB   100.00%    -9223372036854775808
127.0.0.2   rack1      Up     Normal  45.06 KB   100.00%    -3074457345618258603
```

You can see that both the nodes are up and running.

3. Let's start the third node and check the ring status:

```
vivek@vivek-Vostro-3560:~/software/local-cluster/node1$ bin/nodetool ring
```

```
Datacenter: datacenter1
==========
Address     Rack       Status State   Load       Owns       Token
                                                             3074457345618258602
127.0.0.1   rack1      Up     Normal  36.06 KB   66.67%     -9223372036854775808
127.0.0.2   rack1      Up     Normal  45.06 KB   66.67%     -3074457345618258603
127.0.0.3   rack1      Up     Normal  43.7 KB    66.67%     3074457345618258602
```

All three nodes are up and running. With this we end our sample exercise of adding a node on running Cassandra cluster. With Cassandra's ring topology and peer-to-peer architecture, it's really simple to let a node join on the fly. Also, instead of generating the initial token value, we also can set up this with virtual nodes.

Prior to the Cassandra 1.2 release, we needed to manually assign one token range to each data node using the token-generator tool, but with virtual nodes now we can assign multiple token ranges per node. Using the num_tokens property, we can assign a number of virtual nodes per data node, which means holding a large number of smaller data ranges. This is very handy when we need to replace or rebuild a dead node.

A dead node means a node that becomes unresponsive to gossip/intercommunication with other nodes. Such a failure may happen because of network or hardware failures. Each node can be a primary or replica node for data distributed across the Cassandra cluster. When a dead node is ready to join a Cassandra ring, it must stream all data from replica nodes. Without virtual nodes, the contiguous token assignment introduces a lot of overhead as data needs to stream from various primary and replica nodes. With virtual nodes, more tokens assigned to each node mean smaller and more heterogeneous data distribution, which means the same volume of data can be streamed in from multiple nodes! Obviously this would be much faster.

Let's explore replacing a dead node with a new one in a sample exercise.

Replacing a Dead Node

Sometimes it's possible to have a few dead nodes in a cluster. That is, they are not responding to gossip and need to be replaced with new ones. Replacing a dead node in a running Cassandra cluster is straightforward. Let's explore it with a sample exercise continued from the previous one.

1. First, let's bring down node 2 and check the ring status:

```
vivek@vivek-Vostro-3560:~/software/local-cluster/node1$ bin/nodetool ring
```

```
Datacenter: datacenter1
==========
Address      Rack       Status State   Load        Owns       Token
                                                               3074457345618258602
127.0.0.1    rack1      Up     Normal  50.44 KB    66.67%     -9223372036854775808
127.0.0.2    rack1      Down   Normal  54.82 KB    66.67%     -3074457345618258603
127.0.0.3    rack1      Up     Normal  71.09 KB    66.67%     3074457345618258602
```

You can see that node 2 (127.0.0.2) has been down and displayed with Status as Down.

2. Let's configure a new node and assign it a token value that decreases the token at 127.0.0.2 by one (i.e., -3074457345618258604). Before starting the new node, however, let's run a few commands to repair and remove the dead node:

```
vivek@vivek-Vostro-3560:~/software/local-cluster/node1$ bin/nodetool rebuild
vivek@vivek-Vostro-3560:~/software/local-cluster/node1$ bin/nodetool removenode
fd733376-0cf0-49af-bd0b-79ab526350f8
```

Here, fd733376-0cf0-49af-bd0b-79ab526350f8 is the host ID for node 2. You can fetch the host ID for each node by running the nodetool status command.

3. Finally, start the new node configured in the previous step and check the ring status:

```
vivek@vivek-Vostro-3560:~/software/local-cluster/node1$ bin/nodetool status
Datacenter: datacenter1
=======================
Status=Up/Down
|/ State=Normal/Leaving/Joining/Moving
--  Address    Load      Tokens  Owns (effective)  Host ID                                Rack
UN  127.0.0.4  78.66 KB  256     66.7%             cdcecd4b-ac26-4763-99c4-9f757c2d8074   rack1
UN  127.0.0.1  50.44 KB  1       66.7%             8718be22-ed20-43ca-95c1-701864bb1e20   rack1
UN  127.0.0.3  71.09 KB  1       66.7%             0039bf7e-4e9c-4028-b3b6-1d027aedc690   rack1
```

Here, you can see that node 4 (127.0.0.4) is up and joined with the cluster ring.

You must be wondering what happened to node 2 (127.0.0.2)! Once you get node 2 up and running, it automatically replaces the substitute node (i.e., node 4). Whenever node 2 is up and joins the ring, it takes back ownership from node 4. Check node 2's server logs for information (see Figure 9-1)

```
 WARN 15:42:23,301 Token -3074457345618258604 changing ownership from /127.0.0
.4 to /127.0.0.2
 INFO 15:42:23,326 Node /127.0.0.2 state jump to normal
 INFO 15:42:23,330 Startup completed! Now serving reads.
```

Figure 9-1. *Taking ownership back from replacement node (i.e., 127.0.0.4)*

Data Backup and Restoration

Database backup means regularly keeping a copy in a safe location. In case of a natural calamity or hardware loss, the backup can be used to restore the database. Because database scalability and performance should never be at the cost of data loss, Cassandra also provides support for backups and restoration.

Backing up data with Cassandra can be achieved by creating a snapshot. Cassandra provides a mechanism for collecting data snapshots using the nodetool utility. We have already discussed this utility in the current and previous chapters. (Chapter 10 will cover nodetool and other Cassandra-related utilities in detail.) A snapshot of an entire keyspace can be taken when the cluster is up and running. But restoration is possible only by taking the cluster node down.

Using nodetool snapshot and sstableloader

Issuing a snapshot command flushes out all the memtable data and copies it on to the disk and then prepares the hard link with flushed sstables.

Let's discuss this with a sample exercise.

1.   First, let's prepare the schema and populate some data:

```
create keyspace sample_backup with replication = { 'class':'SimpleStrategy','replication_factor':2};
// create keyspace

use sample_backup ; // set keyspace

// create table
create table twitter(hashtag timeuuid primary key, account_name text, tweets set<text>);

// insert records
insert into twitter(hashtag,account_name,tweets) values(now(),'apress_pub',{'New book on
Cassandra is out'});

insert into twitter(hashtag,account_name,tweets) values(now(),'thenews',{'Obama meeting with
European allies on Ukraine http://t.co/l8P8bPcbom from #APress','Bergdahl uproar halts plan for
return celebration http://t.co/BhF6kMy5pW from #APress'});
```

2.   Verify the records have been persisted successfully with the select command:

```
select * from twitter;
 hashtag                              | account_name | tweets
--------------------------------------+--------------+-----------------------------------------
--------------------------------------+--------------+-----------------------------------------
 12da45f0-2702-11e4-861b-9d03f52e8a2b |   apress_pub | {'New book on Cassandra is out'}
 74dc73e0-2702-11e4-861b-9d03f52e8a2b |      thenews | {'Bergdahl uproar halts plan for return
celebration http://t.co/BhF6kMy5pW from #APress', 'Obama meeting with European allies on Ukraine
http://t.co/l8P8bPcbom from #APress'}

(2 rows)
```

3.   To take a snapshot, we need to use the nodetool utility:

```
nodetool -h localhost -p 7199 snapshot twitter

Requested creating snapshot for: twitter
Snapshot directory: 1408816064473
```

The output shows that running nodetool snapshot over a local node has created a snapshot 1408816064473 under the $CASSANDRA_DATA_DIR/twitter/users folder. Here $CASSANDRA_DATA_DIR is the value defined in Cassandra.yaml file for data_file_directories properties (see Figure 9-2).

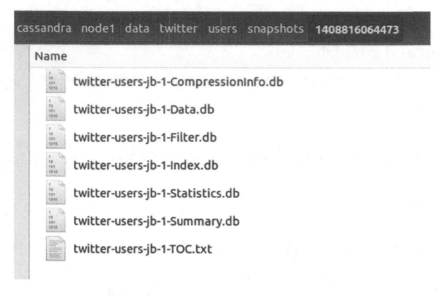

Figure 9-2. The snapshot directory 1408816064473 under the data directory

4. To delete some data that we can restore, let's first truncate the users table:

```
truncate users;
```

5. With step 4, we have truncated the data, so let's try to invoke data restoration. With Cassandra there are multiple ways to initiate restoration. First, let's explore restoration using the sstableloader utility. To begin, we need to copy all .db files in the Snapshot directory into a folder which should be in sync with the database schema, meaning keyspace/tablename. Here in our case it should be the users folder users under twitter (/home/vivek/twitter/users).

6. Next, let's execute the sstableloader utility for restoration:

```
$CASSANDRA_HOME/bin/sstableloader -d localhost  /home/vivek/twitter/users
```

Here /home/vivek/twitter/users is the local folder containing all the backed up .db files. One point worth mentioning is that this snapshot data can also be used for batch analytics using Hadoop MapReduce jobs. You can plug in a custom record reader implementation using org.apache.cassandra.db.OnDiskAtom and org.apache.cassandra.io.sstable.SSTableIdentityIterator. Upon running the preceding command you see the following output on the console:

```
Established connection to initial hosts
Opening sstables and calculating sections to stream
Streaming relevant part of /home/vivek/vivek/twitter/users/twitter-users-jb-1-Data.db to
[/127.0.0.1, /127.0.0.2, /127.0.0.3]
progress: [/127.0.0.2 1/1 (100%)] [/127.0.0.3 1/1 (100%)] [total: 100% - 0MB/s (avg: 0MB/s)]
```

7. Once it completes, we can verify whether the data has been restored by running the select command:

```
select * from twitter.users;

 user_id | age | name
---------+-----+-------
 ckbrown |  32 | chris
  mevivs |  34 | vivek
```

Using nodetool refresh

Another way to restore the data is by using the nodetool refresh utility. It differs from the sstableloader approach in that you need to manually copy .db files in Cassandra data directory. Figure 9-3 shows the structure of the copied sstables.

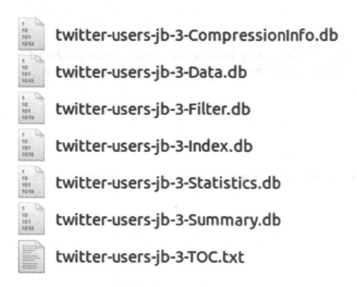

twitter-users-jb-3-CompressionInfo.db

twitter-users-jb-3-Data.db

twitter-users-jb-3-Filter.db

twitter-users-jb-3-Index.db

twitter-users-jb-3-Statistics.db

twitter-users-jb-3-Summary.db

twitter-users-jb-3-TOC.txt

Figure 9-3. *Copied sstables in the Cassandra data directory*

Then run the nodetool refresh command:

```
vivek@vivek-Vostro-3560:~$ $CASSANDRA_HOME/bin/nodetool refresh twitter users
```

Here parameters passed with the nodetool refresh command are the keyspace followed by the table name. This refreshes the data, which can be verified by running the select command.

Using clearsnapshot

Taking snapshots periodically for backup purposes consumes a lot of space on disk; hence, it also requires that you clear obsolete ones from time to time. We can clear the snapshot directory periodically by using `nodetool clearsnapshot` command as follows:

```
vivek@vivek-Vostro-3560:~$ $CASSANDRA_HOME/bin/nodetool -h localhost -p 7199 clearsnapshot
```

The preceding command will clear snapshot directories from all keyspaces. We can also clear snapshot directories for one or more keyspaces like this:

```
/home/vivek/software/local-cluster/node1/bin/nodetool -h localhost -p 7199 clearsnapshot twitter
anotherkeyspace
```

Here, `twitter` and `anotherkeyspace` are selected keyspaces to clear snapshot directories.

In this section, we saw how we can use various utilities to perform database backup and restoration processes. Next we will discuss various monitoring tools.

Cassandra Monitoring Tools

System monitoring is a process of analyzing and gathering the state of the system. Although Cassandra is distributed and fault-tolerant, it also depends on how it is used to build the application. A few key issues which must be taken care periodically are

- Performance
- Data access
- Monitoring node's state

There are many monitoring tools to analyze data and Cassandra's characteristics. A few of the popular ones are

- Helenos
- JConsole
- nodetool
- DataStax DevCenter and OpsCenter

We have already discussed how to monitor and perform logging using JConsole in Chapter 1 (please refer to the "Managing Logs via JConsole" section). Chapter 10 covers more about various Cassandra utilities, including `nodetool`. It is primarily used for Cassandra monitoring and including other operations, some of which we have used earlier in this chapter, like `removenode` and `refresh`.

In this section, we will discuss Helenos and the DataStax DevCenter and OpsCenter for general monitoring.

Helenos

Helenos is a web interface distributed for free under Creative Commons Attribution license. This tool can be used for accessing and manipulating the data and schema. A few of the features supported by Helenos are

- Exploring keyspace and column families
- Schema management
- CQL support
- Authorization and authentication

This project is hosted on https://github.com/tomekkup/helenos. Alternatively you can download the bundled distribution from https://sourceforge.net/projects/helenos-gui/files/.

From above link, you need to download and extract the helenos-1.4-tomcat7_bundle.zip file. Distribution contains Tomcat 7 bundled with a ready-to-use Helenos WAR file.

You just need to start Tomcat by running the catalina.sh shell file (available in the bin folder). Next, enter localhost:8080 in the browser, which will open the Helenos login page (see Figure 9-4).

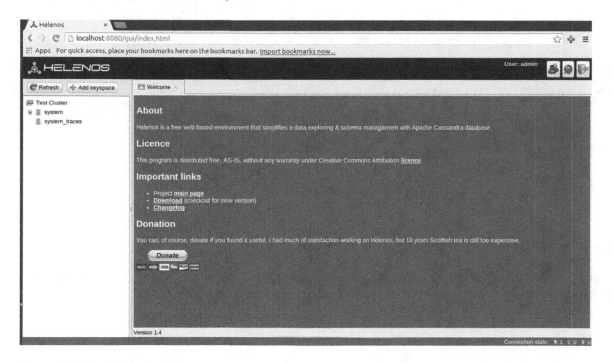

Figure 9-4. *The Helenos login page*

The default credentials are **admin** for Name and **admin** again for Password. After successfully connecting, the console will be displayed as shown in Figure 9-5.

Figure 9-5. *The Helenos console*

In the left panel, you can see a tree view of all the configured keyspaces (see Figure 9-6).

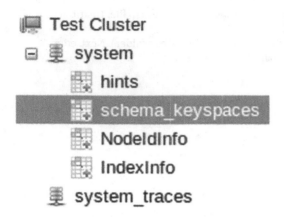

Figure 9-6. *Displaying the cluster and available keyspaces*

We can also explore the data by running CQL3 format queries like the one shown in Figure 9-7.

Name	Value	Clock	TTL	⊡
⊟ ▣ system				
▣ keyspace_name	system	January 1, 1970 05:29:59	0	
▣ durable_writes		March 24, 46599 02:13:05	0	
▣ strategy_class	org.apache.cassandra.locator.LocalStrategy	March 24, 46599 02:13:05	0	
▣ strategy_options	{}	March 24, 46599 02:13:05	0	
⊟ ▣ testkeyspace				
▣ keyspace_name	testkeyspace	January 1, 1970 05:29:59	0	
⊟ ▣ system_traces				
▣ keyspace_name	system_traces	January 1, 1970 05:29:59	0	
▣ durable_writes		January 1, 1970 05:30:00	0	
▣ strategy_class	org.apache.cassandra.locator.SimpleStrategy	January 1, 1970 05:30:00	0	
▣ strategy_options	{"replication_factor":"2"}	January 1, 1970 05:30:00	0	

Figure 9-7. *The result of fetching 100 rows from schema_keyspace table*

The option to add another keyspace or refresh the available schemas using Helenos is available in the top-right corner (refer to Figure 9-5). Figure 9-8 shows the options to add a keyspace or refresh the schema.

Figure 9-8. *We can also add a keyspace or refresh the schema using Helenos*

Adding a table/column family or dropping a keyspace is also possible with Helenos (see Figure 9-9).

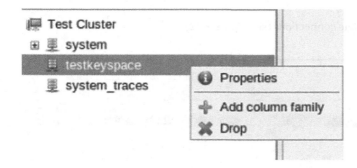

Figure 9-9. *You can add a column family or drop the testkeyspace keyspace*

Selecting the Add column family option opens the dialog box shown in Figure 9-10.

Create new column family

Name * :	twitter
Column * :	Standard
Comparator * :	UTF-8
Subcomparator :	Ascii
Key validation class * :	UTF8Type
Default validation class * :	UTF8Type
GC grace seconds * :	86400
Comment :	Twitter column family

✔ OK   ⊘ Cancel

Figure 9-10. *Adding the Twitter column family*

There is an option for managing connections (see Figure 9-11).

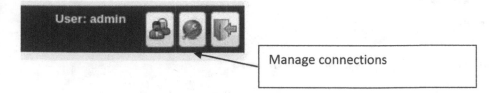

Figure 9-11. *Manage connections with Helenos*

After clicking this option, we can add, edit, or delete connections (see Figure 9-12).

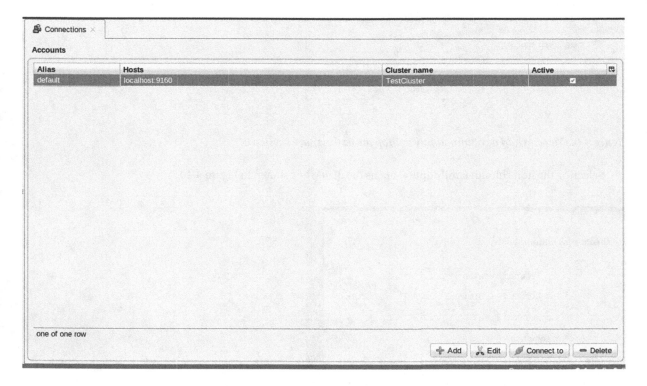

Figure 9-12. *Editing connections with Helenos*

You can also explore pagination, CQL support, and authentication with Helenos. There is a list of planned features at https://github.com/tomekkup/helenos#wish-list, such as exporting data to a file, querying log files, and JMX and cluster monitoring. However, currently there is no active development occurring with Helenos.

DataStax DevCenter and OpsCenter

DataStax is a leading software solution provider company offering commercial products and technical support on Cassandra and the related technology stack. The company's co-founder Jonathan Ellis is also the project chair for the Apache Cassandra project. DataStax offers Datastax Enterprise which comes with DataStax DevCenter and OpsCenter. The company also offers the open source DataStax Java driver for building applications. In this section we will discuss configuring and using these tools for monitoring purposes.

These packages are available for download individually or as part of the DataStax All-in-One Installer at www.datastax.com/download. Whether you download them together or as separate distributions depends on your usage.

OpsCenter

The DataStax OpsCenter can be downloaded as part of the All-in-One Installer package as noted previously or you can download it separately from http://downloads.datastax.com/community/opscenter.tar.gz. The latest release version at the time of writing is 5.0. Once you extract and run opscenter.sh (under the bin folder), the DataStax agents and web interface. The OpsCenter URL is http://localhost:8888. Once you go to this URL in a browser, you see the interface shown in Figure 9-13.

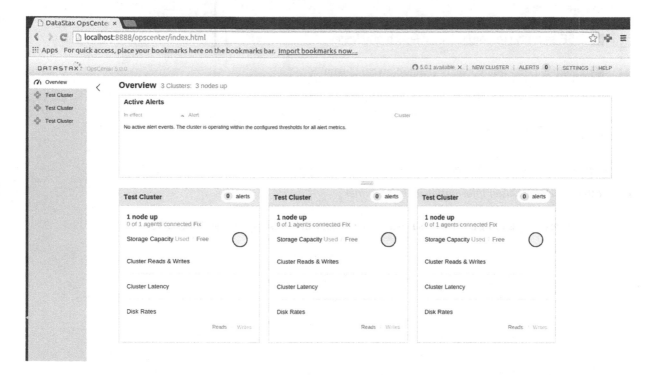

Figure 9-13. *The DataStax OpsCenter console*

The following is a list of features provided by OpsCenter:

- Cluster management. Its health, capacity, and monitoring

- Node and token ring management

- Alert and report generation

In this section we will explore these features. Let's start with cluster management.

We can add a cluster with OpsCenter by clicking the NEW CLUSTER option in the upper-right corner of the console (refer to Figure 9-13). Figure 9-14 shows to the Add Cluster dialog box.

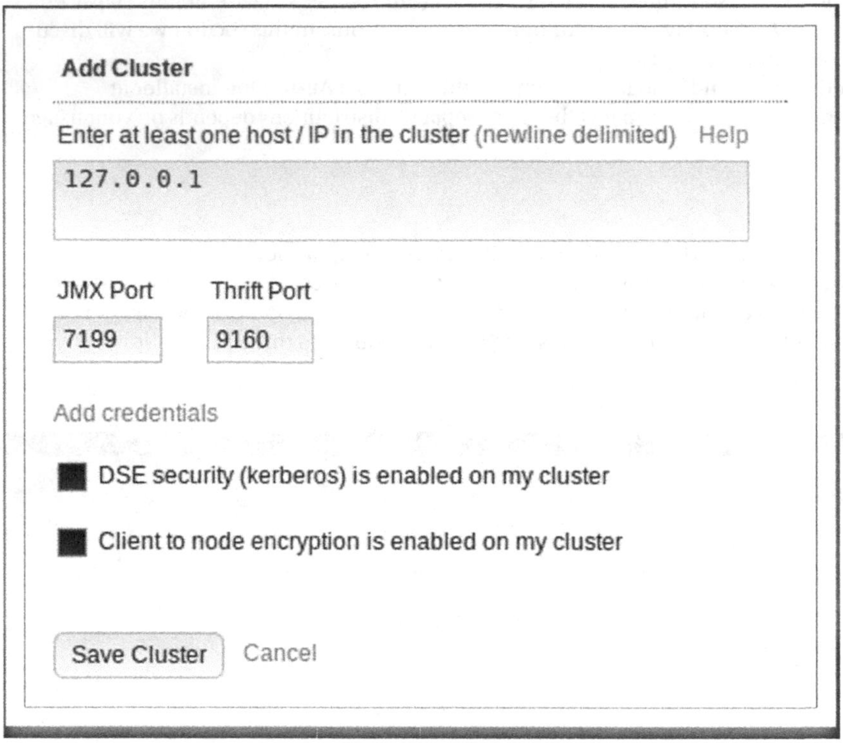

Add Cluster

Enter at least one host / IP in the cluster (newline delimited)   Help

127.0.0.1

JMX Port   Thrift Port

7199   9160

Add credentials

■ DSE security (kerberos) is enabled on my cluster

■ Client to node encryption is enabled on my cluster

Save Cluster | Cancel

Figure 9-14. *Adding cluster information with OpsCenter*

After successfully adding a cluster, its Dashboard screen will be displayed (see Figure 9-15).

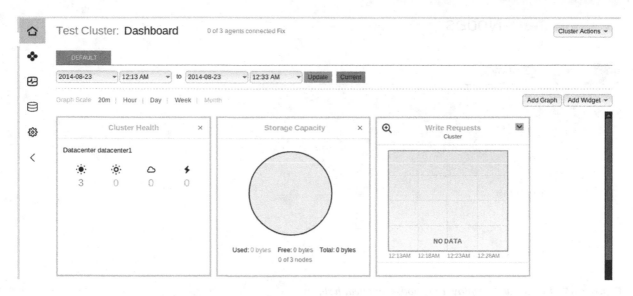

Figure 9-15. *The OpsCenter Dashboard with the cluster named Test Cluster*

By clicking the Test cluster label (refer to Figure 9-13), we can view the cluster ring and nodes status with OpsCenter (shown in Figures 9-16 and 9-17).

DATACENTER ▾	▲ HOSTNAME ▾	TOKEN ▾	STATUS ▾	LOAD (CPU) ▾ Low	DATA SIZE
datacenter1	127.0.0.1	-9223372036854775808	Active	-	
datacenter1	127.0.0.2	-3074457345618258603	Active	-	
datacenter1	127.0.0.3	3074457345618258602	Active	-	

Figure 9-16. *Displaying nodes information and status in the cluster*

Test Cluster: **Nodes**

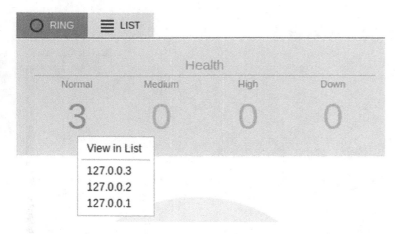

Figure 9-17. *Ring view to display list of nodes and health status*

When you click a node (e.g., 127.0.0.1) from View in List as shown in Figure 9-17, it will show a tab on the Dashboard that enables you to monitor performed events by using the Event logging option (see Figure 9-18).

Time	Level	Message
8/23/2014, 12:33am	Info	Stopping repair service
8/23/2014, 12:33am	Info	OpsCenter starting up.
8/23/2014, 12:31am	Info	Stopping repair service
8/23/2014, 12:31am	Info	Stopping repair service
8/23/2014, 12:31am	Info	OpsCenter starting up.
8/23/2014, 12:31am	Info	OpsCenter starting up.
8/23/2014, 12:31am	Info	Stopping repair service
8/23/2014, 12:31am	Info	OpsCenter shutting down.
8/23/2014, 12:31am	Info	Stopping repair service
8/23/2014, 12:31am	Info	OpsCenter shutting down.
8/22/2014, 5:34pm	Info	Stopping repair service
8/22/2014, 5:34pm	Info	OpsCenter starting up.
8/22/2014, 4:59pm	Info	Stopping repair service
8/22/2014, 4:59pm	Info	OpsCenter starting up.

Figure 9-18. *OpsCenter displays event logging with event messages*

The OpsCenter Enterprise offering provides features such as backup and restoration, node rebalancing, and so forth. You can experiment with OpsCenter using the community edition by adding widgets, such as a graph widget, for exporting and managing data in a graphical format. You can also refer to www.datastax.com/what-we-offer/products-services/datastax-opscenter/compare for information on community vs. enterprise offerings of DataStax OpsCenter.

DevCenter

DataStax DevCenter is a visual interface for managing schema and performing CQL3 queries. It comes in very handy for quick data operations over Cassandra. The latest available version is 1.2. After downloading (available in the All-in-One Installer at www.datastax.com/download or individually at www.datastax.com/download#dl-devcenter) and extracting it successfully, we can start by running devcenter.sh (under the bin folder). That would open the DevCenter Connection Manager console to enter cluster connection details, as shown in Figure 9-19.

DevCenter Connection Manager

Create a New Connection

Connection name: `local-cluster`

Contact hosts: | Add

127.0.0.1 | Remove

Native Protocol port: `9042` | Test

Use compression: ● None ○ Snappy ○ LZ4

Figure 9-19. The DevCenter Connection Manager console

After successfully connecting, the DevCenter workbench interface displays (see Figure 9-20).

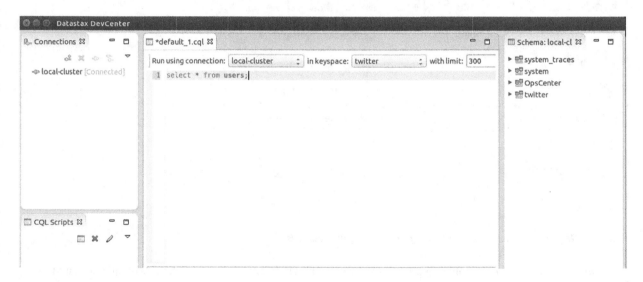

Figure 9-20. Datastax DevCenter with the query editor

You can execute a CQL3 query using ALT+F11, and the output will be displayed as shown in Figure 9-21.

Results		
user_id	age	name
ckbrown	32	chris
mevivs	34	vivek

Figure 9-21. *Query output in the DevCenter workbench*

DevCenter is a very useful tool for developers and admins for creating and managing schema. You can also use it to explore the various data modeling techniques discussed in Chapters 2 and 3.

The DataStax OpsCenter and DevCenter user guides are available at www.datastax.com/documentation/opscenter/5.0/opsc/about_c.html and www.datastax.com/documentation/getting_started/doc/developer/devcenter/usingDevCenter.html for reference. These are commercial offerings, so a few of the features may only be available with the Enterprise edition. The Community edition is available for free download but for limited purpose usage. More details around the licensing model are available at www.datastax.com/download/dse-vs-dsc and www.datastax.com/developer-license-terms.

Summary

In this chapter we discussed various tools available for Cassandra monitoring and data manipulation. This chapter also covered basics of data backup and restoration along with managing nodes in a cluster. The next chapter will discuss various Cassandra utilities and performing benchmarking using YCSB.

CHAPTER 10

■ ■ ■

Cassandra Utilities

So far in this book, we have covered almost all aspects of Cassandra, including data modeling, graph databases, batch analytics, performance tuning, and analyzing various open source and commercial tools. Since we are progressing toward the end of this book, one area worth discussing is various built-in utilities which are quite useful for database administration, stress testing, or performing data migration or bulk loading.

Some of these topics have been discussed already in this book. For example, the stress testing tool for command line-based stress testing was described in Chapter 8, the token generator for token assignment in Chapter 1, and a few nodetool options in Chapters 3, 4, and 5.

In this chapter we will solely discuss more Cassandra utilities, including:

- The nodetool utility

- sstable2json and json2sstable

- Cassandra bulk loading (with CQL3 collections)

Cassandra nodetool Utility

Cassandra provides a number of operations, such as token assignment, ring management, schema management, node management and monitoring, and so forth. Cassandra's nodetool utility is a command-line interface that wraps all such operations with simple commands. In this section, we will explore a number of these commands in detail.

The nodetool utility is built-in and provided with the Cassandra distribution (found in the $CASSANDRA_HOME/bin folder).

Before moving to the commands, let's print the nodetool version:

```
vivek@vivek-Vostro-3560:~/software/apache-cassandra-2.0.4/bin$./nodetool version
ReleaseVersion: 2.0.4
```

The output from the preceding command shows that nodetool's release version is 2.0.4.

To print all the available commands with this utility, simply run the following command:

```
vivek@vivek-Vostro-3560:~/software/apache-cassandra-2.0.4/bin$./nodetool
```

Let's start our discussion of the nodetool utility commands with Cassandra ring management.

Ring Management

A Cassandra ring is made up of various data nodes. Two of the important operations from a ring management perspective are

- Checking ring status

- Decommissioning a dead node

Checking Ring Status

In Chapter 1, in the "Configuring Multiple Nodes over a Single Machine" section, we discussed how to start a Cassandra cluster on a local machine. Refer to Chapter 1 for information on starting a Cassandra cluster over a single machine. Once the cluster has been started, we can print the Cassandra cluster ring information like this:

```
vivek@vivek-Vostro-3560:~/software/local-cluster/node4$ bin/nodetool ring
```

Running the preceding command prints this output on the console:

```
Datacenter: datacenter1
==========
Address      Rack     Status   State     Load         Owns       Token
127.0.0.4    rack1    Up       Normal    77.55 KB     66.67%     -3074457345618258604
127.0.0.1    rack1    Up       Normal    35.79 KB     66.67%     -9223372036854775808
127.0.0.2    rack1    Up       Normal    45.05 KB     33.33%     -3074457345618258603
127.0.0.3    rack1    Up       Normal    61.28 KB     33.33%      3074457345618258602
```

Here the ring command prints each node's status and the assigned tokens for each Cassandra node. The output shows the data nodes' IP addresses, their statuses, and rack information:

- The Address column shows the nodes' URL or IP address.

- The Rack column contains rack information.

- If the nodes are live and are part of the cluster ring, their status will be Up; otherwise, it will be Down.

- Possible values for State are Normal, Leaving, Joining, and Moving.

  - If a data node is part of cluster ring it will be shown with status as Normal.

  - If a node is getting decommissioned, then it will have an intermediate state of Leaving.

  - The status will be Moving if the node has been moved to another token value.

- The Load column displays the current data load on each node. This information is important from an administration perspective as load and ownership (the Owns column) information show whether the ring is in a balanced state.

- The Token column shows the assigned token, which is a token range.

To fetch more information on the Cassandra cluster, like current state, load, and assigned token, we can run the status command:

```
vivek@vivek-Vostro-3560:~/software/local-cluster/node1$ bin/nodetool status
```

Running this command prints the data shown in following output:

```
Datacenter: datacenter1
=======================
Status=Up(U)/Down(D)
|/ State=Normal(N)/Leaving(L)/Joining(J)/Moving(M)

Status/  Address    Load      Tokens  Owns        Host ID                                Rack
State                                 (effective)
UN       127.0.0.1  59.82 KB  129     166.7%      718d384e-c8c2-4355-bf9b-1bfa60be2d99   Rack1
UN       127.0.0.2  66.54 KB  256     166.7%      cb9da211-6948-4c21-b5f5-5ee660cd9355   Rack1
UN       127.0.0.3  50.31 KB  256     66.7%       74ee5df6-d990-4deb-9144-501c349308d6   Rack1
```

The preceding output is similar to the first output example in this section. The only differences are the Tokens and Host ID fields. Here the Tokens column shows the number of tokens set for each node, and the Host ID is the network ID of the data node. It's a UUID value assigned by Cassandra for each node.

That's how we can monitor the cluster ring status and its information. In the next section we will discuss how to decommission a live data node.

Decommissioning a Node

We can also decommission a node using the nodetool utility. In this exercise, we will decommission a node by using the decommission command and then verify that it is decommissioned by checking the cluster ring status.

1. Assuming all three nodes are up and running (see the output shown in the preceding section), let's create a keyspace **twitter_keyspace** and table **twitter** using the cql command-line interface:

   ```
   cqlsh> create keyspace twitter_keyspace with replication =
   {'class':'SimpleStrategy', 'replication_factor':3};
   cqlsh> use twitter_keyspace;

   cqlsh:twitter_keyspace> create table twitter(tweeted_at text, screen_name text,
   body text, PRIMARY KEY (tweeted_at,screen_name));
   ```

2. Let's load data in the twitter table using the copy command:

   ```
   cqlsh:twitter_keyspace> copy twitter(tweeted_at,screen_name,body) from
   '/home/vivek/tweets_pipe' WITH DELIMITER = '|';
   10020 rows imported in 12.457 seconds.
   ```

Here /home/vivek/tweets_pipe is a local file in which all column values have a "|" field separator. You can find this file with downloads for this book, or you can create the sample using the following data and format:

```
2014-08-02|mevivs|Working on Cassandra book
2014-08-04|chris_nelson|Cassandra book review in progress
```

Now let's try to decommission the node with the IP address 127.0.0.1 using the nodetool utility's decommission command:

```
vivek@vivek-Vostro-3560:~/software/local-cluster/node1$ bin/nodetool decommission
```

Please note that we can only decommission a live node. Decommissioning a node would implicitly start data streaming to another node. And we can monitor the decommissioning process using the nodetoolnetstats command:

```
vivek@vivek-Vostro-3560:~/software/local-cluster/node1$ bin/nodetool netstats
```

It would display the following output:

```
Mode: LEAVING
Not sending any streams.
Read Repair Statistics:
Attempted: 0
Mismatch (Blocking): 0
Mismatch (Background): 0
Pool Name                Active    Pending      Completed
Commands                    n/a          0          20048
Responses                   n/a          0           4806
```

The output means the node is still being processed and is in the LEAVING state. Here Attempted means the number of attempts for the read repair. Mismatch (Blocking) and Mismatch (Background) refer to the number of read repairs since the server restart that blocked query and background server restart, respectively. Here Commands holds information about the number of read/write operations, and Responses holds information about the received responses on client read/write requests.

During the LEAVING state, if we check cluster status, it will display output as shown after the command:

```
vivek@vivek-Vostro-3560:~/software/local-cluster/node1$ bin/nodetool status
```

```
Datacenter: datacenter1
========================
Status=Up(U)/Down(D)
|/ State=Normal(N)/Leaving(L)/Joining(J)/Moving(M)

Status/  Address    Load       Tokens  Owns       Host ID                                Rack
State                                  (effective)
UL       127.0.0.1  59.82 KB   129     166.7%     718d384e-c8c2-4355-bf9b-1bfa60be2d99   Rack1
UN       127.0.0.2  66.54 KB   256     166.7%     cb9da211-6948-4c21-b5f5-5ee660cd9355   Rack1
UN       127.0.0.3  50.31 KB   256     66.7%      74ee5df6-d990-4deb-9144-501c349308d6   Rack1
```

Once decommissioning of node has been completed, you can validate it using the netstat command, and the output will be as follows:

```
vivek@vivek-Vostro-3560:~/software/local-cluster/node1$ bin/nodetool netstats
```

```
Mode: DECOMMISSIONED
Not sending any streams.
Read Repair Statistics:
Attempted: 0
Mismatch (Blocking): 0
Mismatch (Background): 0
```

```
Pool Name              Active    Pending    Completed
Commands                 n/a        0         20048
Responses                n/a        0          4808
```

Here you can see that mode is now DECOMMISSIONED.

3. Once the node has been decommissioned, we can validate it using the status command.

vivek@vivek-Vostro-3560:~/software/local-cluster/node1$ bin/nodetool status

```
Datacenter: datacenter1
=======================
Status=Up(U)/Down(D)
|/ State=Normal(N)/Leaving(L)/Joining(J)/Moving(M)

Status/  Address    Load      Tokens  Owns       Host ID                                Rack
State                                 (effective)
UN       127.0.0.2  66.54 KB  256     100%       cb9da211-6948-4c21-b5f5-5ee660cd9355   Rack1
UN       127.0.0.3  50.31 KB  256     100%       74ee5df6-d990-4deb-9144-501c349308d6   Rack1
```

Schema Management

Schema management means analyzing keyspaces and tables, performing repairs, rebuilding or cleaning up data, and various other activities. With the nodetool utility, these operations are available in the form of simple commands. Using them we can easily manage schema level operations. Let's discuss some of the more important ones with simple examples.

cfstats

Using the cfstats command we can gather statistics about keyspaces and their tables. It is very useful to monitor read/write latency, compaction, and memtable-related information.

We can collect information for a keyspace like this:

vivek@vivek-Vostro-3560:~/software/local-cluster/node4$ bin/nodetool cfstats twitter_keyspace

Here **twitter_keyspace** is the keyspace name.
We can also collect statistics for multiple keyspaces:

vivek@vivek-Vostro-3560:~/software/local-cluster/node4$ bin/nodetool cfstats keyspace1 keyspace2

Here **keyspace1** and **keyspace2** are those keyspaces.
Also, we can capture statistics for a specific table:

vivek@vivek-Vostro-3560:~/software/local-cluster/node4$ bin/nodetool cfstats twitter_keyspace.twiiter

Here, **twitter_keyspace** is the keyspace and **twitter** is the table.

The following is the output from running cfstats against the **twitter** table:

```
Keyspace: twitter_keyspace
    Read Count: 0
    Read Latency: NaN ms.
    Write Count: 1
    Write Latency: 1.314 ms.
    Pending Tasks: 0
        Table: twitter
        SSTable count: 0
        Space used (live), bytes: 0
        Space used (total), bytes: 0
        SSTable Compression Ratio: 0.0
        Number of keys (estimate): 0
        Memtable cell count: 3
        Memtable data size, bytes: 590
        Memtable switch count: 0
        Local read count: 0
        Local read latency: NaN ms
        Local write count: 1
        Local write latency: 1.314 ms
        Pending tasks: 0
        Bloom filter false positives: 0
        Bloom filter false ratio: 0.00000
        Bloom filter space used, bytes: 0
        Compacted partition minimum bytes: 0
        Compacted partition maximum bytes: 0
        Compacted partition mean bytes: 0
        Average live cells per slice (last five minutes): 0.0
        Average tombstones per slice (last five minutes): 0.0
```

cfhistogram

The cfhistogram command provides statistics about read/write latency in microseconds and the number of sstables involved during reads and writes. The following command shows output of a column family histogram of the **twitter** table within the keyspace **twitter_keyspace**:

```
vivek@vivek-Vostro-3560:~/software/local-cluster/node4$ bin/nodetool cfhistograms
twitter_keyspace twiiter
twitter_keyspace/twiiter histograms
```

Offset	SSTables	Write Latency (micros)	Read Latency (micros)	Partition size (bytes)	Cell Count
1	0	0	0	0	0
2	0	0	0	0	0
3	234	0	0	0	0
4	0	0	0	0	0
5	0	0	0	0	0
6	0	0	0	0	0
7	0	0	0	0	0

8	0	0	0	0	0
10	0	0	0	0	0
12	0	0	0	0	0
14	0	0	0	0	0
17	0	0	0	0	0
20	0	0	0	0	0
24	0	0	0	0	0
29	0	127	86	0	0
35	0	149	0	0	0
42	0	0	0	0	0
50	0	0	0	0	0
60	0	0	0	0	0

Here, the third row in the output table depicts that there is a total of 3 sstable lookups done to serve 234 read requests. The value in the SSTables column denotes the number of read requests. Also the row having offset 29 denotes read latency for 86 read requests and 127 write requests is between 24 to 29 microseconds, whereas for 149 write request write latency is between 29 and 35 microseconds.

This comes in very handy when we need to analyze a table's performance under heavy read or write loads.

cleanup

We can run the cleanup command to clean up keyspaces and partition keys which are not related to data nodes, or data that doesn't belong to that node.

Generally we execute this command to delete data from a node that has just been added to the cluster. Whenever a new node joins in the ring, using the cleanup command we can remove the obsolete data that is not related to the cluster. Also we perform this command whenever a node becomes live after it has been decommissioned or removed from the cluster ring. We can run this command:

```
vivek@vivek-Vostro-3560:~/software/local-cluster/node4$bin/nodetool cleanup twitter_keyspace twiiter
```

Here we are performing cleanup on **twitter** table of keyspace **twitter_keyspace**. Please note that if we don't specify the keyspace, it will perform a cleanup over all available keyspaces. Also, we can perform a cleanup over multiple tables in a keyspace by providing a list of table names:

```
nodetool cleanup   <keyspace><table1><table2>
```

clearsnapshot

Snapshot directories are generally created for the backup and restore processes. When snapshot directories are no longer required for a keyspace, we can clear them using the clearsnapshot command:

```
vivek@vivek-Vostro-3560:~/software/local-cluster/node4$ bin/nodetool clearsnapshot twitter_keyspace
```

Running the preceding command, deletes all snapshot directories for the keyspace **twitter_keyspace**. We can also delete specific snapshot directories using the -f option as follows:

```
vivek@vivek-Vostro-3560:~/software/local-cluster/node4$ bin/nodetool clearsnapshot
twitter_keyspace -f 1412341689
```

Generally, clearing snapshots is part of the backup process, and regular backing up and deleting snapshot directories would be scheduled background cron jobs. A cron job is a time-based scheduler that can be set up as a component of the backup process.

■ **Note**   For more information on data backup and restore, please refer to the "Data Backup and Restoration" section in Chapter 9.

---

flush

A memtable is a cache of data rows that can be looked up for a particular row key and data. An implicit flush of memtables is automatically handled by Cassandra based on the memtable threshold and throughput configurations defined in `cassandra.yaml`. Using the `nodetool flush` command, we can explicitly flush data on a particular data node. Reasons for explicit flushes include

- The node is running out of space and needs to free some space on the server side.

- Data backup is occurring and you might need to restore it on another node using the sstable loader.

- A flush out to disk is required to reduce commit log replay during node restart.

Explicitly flush memtables onto disk using the `nodetool flush` command as follows:

```
vivek@vivek-Vostro-3560:~/software/local-cluster/node4$ bin/nodetool flush twitter_keyspace twitter
```

Running the preceding command explicitly flushes out data from memtables and writes it onto disk for the keyspace **twitter_keyspace** and table **twitter**.

repair

You need to repair cluster nodes generally if

- One or multiple nodes are in recovery mode.

- One of the Cassandra nodes is not participating in reads.

- A node is being added and data needs to manually be updated on that node.

Ideally, performing repairs periodically is recommended to to keep clusters consistent and in balanced states. We can run the `repair` command as follows:

```
vivek@vivek-Vostro-3560:~/software/local-cluster/node4$ bin/nodetool repair -dc DC1
```

Running the preceding command triggers the repair process across multiple nodes on data center DC1. Please note that repair is a memory-intensive process and requires that you perform disk I/O to copy snapshots across replica nodes. We can also perform the repair process in parallel using the -pr option, which initiates the repair process across replica nodes and results in less downtime. We can also restrict the repair process to nodes that are local to a data center like this:

```
vivek@vivek-Vostro-3560:~/software/local-cluster/node4$ bin/nodetool repair -local
```

rebuild

We perform the rebuild command whenever we add a new data center in a Cassandra cluster and need to copy data from one data center to the newly added one. The way to perform this command is

```
vivek@vivek-Vostro-3560:~/software/local-cluster/node4$ bin/nodetool rebuild - - DC2
```

Here DC2 is a new data center. Please make sure to include the newly added data center, or the command will perform successfully but no data copy will happen.

rebuild_index

In Chapter 2, we discussed that during schema changes, most often during Thrift and CQL3 exchange, it is quite possible for a schema and the secondary indexes to be corrupted. We can rebuild those indexes using the rebuild_index command:

```
vivek@vivek-Vostro-3560:~/software/local-cluster/node4$ bin/nodetool rebuild_index
twitter_keyspace twitter
```

Running the preceding command does a full index rebuild on all indexes of the **twitter** table in keyspace **twitter_keyspace**. We can also rebuild specific indexes as follows:

```
vivek@vivek-Vostro-3560:~/software/local-cluster/node4$ bin/nodetool rebuild_index
twitter_keyspace twitter  screenname_idx
```

Here screenname_idx is the index on column screen_name.

Rebuilding indexes will not drop the index, but it will rebuild it with the available data on that node. There are a lot of other command options with the nodetool utility. Here I have covered the most important ones from a data administration backup perspective. In the following sections, we will discuss the refresh, repair, and sstablescrub commands, as well.

JSONifying Data

Since its inception, the Cassandra distribution has come with tools to import JSON-format data into sstable and export sstable data in JSON files. These tools were primarily built for debugging purposes, but they also are useful when we need to migrate data from one node to another.

The two tools that come with the Cassandra distribution are sstabl2json and json2sstable. We can find them in the $CASSANDRA_HOME/bin folder. Let's explore both!

Exporting Data to JSON Files with sstable2json

Using the sstabl2json command we can export on disk sstables data in JSON files. Let's explore it with the following exercise:

1. First let's create the schema and populate data using the cql shell. We will create a keyspace **twitterkeyspace** and table **user**:

    ```
    create keyspace twitterkeyspace with replication = {'class':'SimpleStrategy',
    'replication_factor':3}
    use twitterkeyspace;
    create table user(user_id timeuuid primary key, fname text,lname text);
    ```

2. Let's insert a few records:

```
insert into user(user_id,fname,lname) values(now(),'vivek','mishra');
insert into user(user_id,fname,lname) values(now(),'Melissa','Maldonado');
insert into user(user_id,fname,lname) values(now(),'Chris','Nelson');
insert into user(user_id,fname,lname) values(now(),'Brian','Neill');
select * from user;
 user_id                              | fname   | lname
--------------------------------------+---------+-----------
 849d5820-50e3-11e4-abc1-3f484de45426 |   Brian |    Neill
 74ac3a80-50e3-11e4-abc1-3f484de45426 | Melissa | Maldonado
 7c34a260-50e3-11e4-abc1-3f484de45426 |   Chris |    Nelson
 f7d1f2c0-50e2-11e4-abc1-3f484de45426 |   vivek |    mishra
```

3. Let's flush this data into sstables using nodetool flush:

```
$CASSANDRA_HOME/bin/nodetool flush
```

4. Finally, export data in a JSON file using sstable2json:

```
$CASSANDRA_HOME/bin/sstable2json /var/lib/cassandra/data/twitterkeyspace/user/
twitterkeyspace-user-jb-1-Data.db > output.json
```

Here twitterkeyspace-user-jb-1-Data.db is a .db data file under the data directory (in this case it is under /var/lib/Cassandra). The preceding command will create a JSON file having data as shown in Figure 10-1.

```
[
{"key": "849d582050e311e4abc13f484de45426","columns": [["","",1412989884066000], ["fname","Brian",1412989884066000],
["lname","Neill",1412989884066000]]},
{"key": "74ac3a8050e311e4abc13f484de45426","columns": [["","",1412989857319000], ["fname","Melissa",1412989857319000],
["lname","Maldonado",1412989857319000]]},
{"key": "7c34a26050e311e4abc13f484de45426","columns": [["","",1412989869958000], ["fname","Chris",1412989869958000],
["lname","Nelson",1412989869958000]]},
{"key": "f7d1f2c050e211e4abc13f484de45426","columns": [["","",1412989647852000], ["fname","vivek",1412989647852000],
["lname","mishra",1412989647852000]]}
]
```

Figure 10-1. *Data in the output.json file*

We can also include and exclude specific rows using the –k and –x options, respectively. With the –k option, keys have to be in hexadecimal format. Since with the user table, the primary key is of timeuuid type, so let's first get key values in hexadecimal format using the timeuuidAsBlob function as follows:

```
select timeuuidAsBlob(user_id),fname from user;
 timeuuidAsBlob(user_id)            | fname
------------------------------------+---------
 0x849d582050e311e4abc13f484de45426 |   Brian
 0x74ac3a8050e311e4abc13f484de45426 | Melissa
 0x7c34a26050e311e4abc13f484de45426 |   Chris
 0xf7d1f2c050e211e4abc13f484de45426 |   vivek
(4 rows)
```

Here `timeuuidAsBlob(user_id)` is the `user_id` key in hexadecimal format. Now we can fetch rows having `fname` as brian and vivek using the –k option:

```
$CASSANDRA_HOME/bin/sstable2json /var/lib/cassandra/data/twitterkeyspace/user/twitterkeyspace-user-
jb-1-Data.db -k 849d582050e311e4abc13f484de45426 f7d1f2c050e211e4abc13f484de45426
[
{"key": "849d582050e311e4abc13f484de45426","columns": [["",""],1412989884066000],
["fname","Brian",1412989884066000], ["lname","Neill",1412989884066000]]},
{"key": "f7d1f2c050e211e4abc13f484de45426","columns": [["",""],1412989647852000],
["fname","vivek",1412989647852000], ["lname","mishra",1412989647852000]]}
]
```

849d582050e311e4abc13f484de45426 and f7d1f2c050e211e4abc13f484de45426 are the user IDs for brian and vivek.

Similarly, we can exclude specific keys using the –x option:

```
$CASSANDRA_HOME/bin/sstable2json /var/lib/cassandra/data/twitterkeyspace/user/twitterkeyspace-user-
jb-1-Data.db -x 849d582050e311e4abc13f484de45426 f7d1f2c050e211e4abc13f484de45426
[
{"key": "74ac3a8050e311e4abc13f484de45426","columns": [["",""],1412989857319000],
["fname","Melissa",1412989857319000], ["lname","Maldonado",1412989857319000]]},
{"key": "7c34a26050e311e4abc13f484de45426","columns": [["",""],1412989869958000],
["fname","Chris",1412989869958000], ["lname","Nelson",1412989869958000]]}
]
```

Importing JSON Data with json2sstable

Similarly we can import JSON data back into sstable using `json2sstable` in a few simple steps:

1. First create a table **dumpuser** with the keyspace **twitterkeyspace** using the cql shell:

   ```
   use twitterkeyspace;
   create table dumpuser(user_id timeuuid primary key, fname text,lname text);
   ```

2. Next, import data from `output.json` using `json2sstable` in the **dumpuser** table:

   ```
   $CASSANDRA_HOME/bin/json2sstable -K twitterkeyspace -c dumpuser output.json /var/
   lib/cassandra/data/twitterkeyspace/dumpuser/twitterkeyspace-dumpuser-jb-1-Data.db
   Importing 4 keys...
   4 keys imported successfully.
   ```

3. Next refresh the newly loaded sstables using the `nodetool refresh` command:

   ```
   $CASSANDRA_HOME/bin/nodetool refresh twitterkeyspace dumpuser
   ```

   `nodetool refresh` makes data available without a node restart.

4. Now we can verify the loaded data by running the `select` command using the cql shell over the table **dumpuser**:

```
cqlsh:twitterkeyspace> select * from dumpuser;
 user_id                               | fname   | lname
---------------------------------------+---------+----------
 849d5820-50e3-11e4-abc1-3f484de45426  |  Brian  |   Neill
 74ac3a80-50e3-11e4-abc1-3f484de45426  | Melissa | Maldonado
 7c34a260-50e3-11e4-abc1-3f484de45426  |  Chris  |  Nelson
 f7d1f2c0-50e2-11e4-abc1-3f484de45426  |  vivek  |  mishra
 (4 rows)
```

One piece of advice here is to avoid trying to load data in sstables for the same table using `json2sstable` multiple times, as it may leave an sstable in a corrupted state. You can check `server.log` (under the log directory) for the following error:

```
Caused by: org.apache.cassandra.io.compress.CorruptBlockException: (/var/lib/cassandra/data/
twitterkeyspace/dumpuser/twitterkeyspace-dumpuser-jb-1-Data.db): corruption detected,
chunk at 0 of length 210.
at org.apache.cassandra.io.compress.CompressedRandomAccessReader.decompressChunk
(CompressedRandomAccessReader.java:122)
at org.apache.cassandra.io.compress.CompressedRandomAccessReader.reBuffer
(CompressedRandomAccessReader.java:87)
... 26 more
```

Trying to fetch data from a corrupted table (in this case the **dumpuser** table) using the cql shell results in the following output:

```
cqlsh:twitterkeyspace> select * from dumpuser;
```
Request did not complete within rpc_timeout.

In this situation scrubbing and removing the corrupted data is our only option.

1. The Cassandra distribution comes with `sstablescrub` utility, which we'll use here:

   ```
   $CASSANDRA_HOME/bin/sstablescrub twitterkeyspace dumpuser
   ```

Running the preceding command results in the following output:

```
Pre-scrub sstables snapshotted into snapshot pre-scrub-1412993550669
Scrubbing SSTableReader(path='/var/lib/cassandra/data/twitterkeyspace/dumpuser/twitterkeyspace-
dumpuser-jb-1-Data.db') (218 bytes)
Error scrubbing SSTableReader(path='/var/lib/cassandra/data/twitterkeyspace/dumpuser/
twitterkeyspace-dumpuser-jb-1-Data.db'): null
```

2. As a prestep, it's always better to take a backup before running the `repair` command. First create a snapshot directory for a backup. As you can see in the preceding output, an error occurred while scrubbing the `.db` file; hence, we need to manually remove `/twitterkeyspace-dumpuser-jb-1-Data.db` and restore the rest of the data from the snapshot directory. Please refer to Chapter 9, the "Data Backup and Restoration" section, for how to restore data from the snapshot directory.

3. Also we need to run the `nodetool repair` command to bring back the table:

    ```
    $CASSANDRA_HOME/bin/nodetool repair twitterkeyspace dumpuser
    ```

So here we have seen how to import JSON data into sstables and export data to JSON files, as well how to deal with corrupted sstables.

Currently, there are discussions in the community about whether to retire or replace these tools. You can refer to `https://issues.apache.org/jira/browse/CASSANDRA-7464` for more details.

Cassandra Bulk Loading

In the "Decommissiong a Node" section, we saw that using the `COPY` command we can copy data from a `.csv` file to tables. Prior to CQL3's inception, in the early releases of Cassandra, the distribution came with a tool called sstableloader, which could be used to directly load `.db` files into sstables. We just had to write an implementation using the SSTableWriter API to generate `.db` files. One such Thrift-based implementation can be downloaded from `www.datastax.com/wp-content/uploads/2011/08/DataImportExample.java`. Since CQL3, however, the CQL3 binary protocol is going to be the active protocol for future Cassandra development, and also please note that support for Thrift has been discontinued.

Here I am sharing a CQL3 compatible implementation to generate `.db` files for CQL3 tables. This is the extended implementation of the preceding shared URL.

In this example we will create a table having a column of the collection type. Let's start!

1. First we need to create keyspace `Cql3Demo` and table `Users`:

    ```
    create keyspace "Cql3Demo"  with replication = { 'class' :
    'SimpleStrategy', 'replication_factor':3};
    use "Cql3Demo";
    create table "Users"(user_id uuid PRIMARY KEY,firstname text,lastname text,
    password text,email text, age int, addresses set<text>);
    ```

2. Create a `.csv` file (`CSVInput.csv`) having a user ID, first name, last name, password, email address, and age as follows:

    ```
    5bd8c586-ae44-11e0-97b8-0026b0ea8cd0,vivek,mishra,4Jc2s,22,vivek.mishra@nomail.com
    4bd8cb58-ae44-12e0-a2b8-0026b0ed9cd1,apress,bigdata,s!a0ml,12,bigdata@nomail.com
    1ce7cb58-ae44-12e0-a2b8-0026b0ad21ab,Brian,Neil,s)3B3,12,brian.neil@nomail.com
    ```

3. Now, let's have a look at implementation. It's a Maven-based project and requires that we add the following dependency in `pom.xml`:

    ```
    <dependency>
        <groupId>org.apache.cassandra</groupId>
        <artifactId>cassandra-all</artifactId>
        <version>2.0.4</version>
    </dependency>

    <dependency>
        <groupId>org.apache.cassandra</groupId>
        <artifactId>cassandra-clientutil</artifactId>
    <version>2.0.4</version>
    </dependency>
    ```

4. Now let's have a look at the Java implementation:

```java
if (args.length == 0)
        {
filename = " CSVInput.csv";
        }
else
        {
filename = args[0];
        }
        BufferedReader reader = new BufferedReader(new FileReader(filename));

        String keyspace = "Cql3Demo";

        String columnFamily = "Users";

// create cassandra type structure default is data/KEYSPACE/users

        File directory = new File("data");

if (!directory.exists())
        {
            directory.mkdir();
        }

        directory = new File(directory.getPath() + "/" + keyspace);
if (!directory.exists())
            directory.mkdir();

        directory = new File(directory.getPath() + "/" + columnFamily);
if (!directory.exists())
            directory.mkdir();
```

The preceding code snippet shows that the program takes the file name as an input argument. If it isn't provided, then the default file (CSVInput.csv) is used. It will also create a data folder in the root folder (see Figure 10-2).

- ▲ 📁 data
  - ▲ 📁 Cql3Demo
    - 📁 Users

Figure 10-2. *The created data folder with .db files*

5. Next, we need to instantiate SSTableSimpleUnsortedWriter:

```java
SSTableSimpleUnsortedWriter usersWriter = new SSTableSimpleUnsortedWriter
(directory, new Murmur3Partitioner(),
keyspace, "Users", AsciiType.instance, null, 64, new CompressionParameters(
org.apache.cassandra.io.compress.SnappyCompressor.create(Collections.<String,
String> emptyMap()))));
```

6. Now, we need to parse each entry in the CSVInput.csv file:

```java
static class CsvEntry
{
    UUID key;

    String firstname;

    String lastname;

    String password;

    int age;

    String email;

    boolean parse(String line, int lineNumber)
    {
        // Ghetto csv parsing
        String[] columns = line.split(",");
        if (columns.length != 6)
        {
            System.out.println(String.format("Invalid input '%s' at line %d
            of %s", line, lineNumber, filename));
            return false;
        }
        try
        {
            key = UUID.fromString(columns[0].trim());
            firstname = columns[1].trim();
            lastname = columns[2].trim();
            password = columns[3].trim();
            age = Integer.parseInt(columns[4].trim());
            email = columns[5].trim();
            return true;
        }
        catch (NumberFormatException e)
        {
            System.out.println(String.format("Invalid number in input '%s'
            at line %d of %s", line, lineNumber, filename));
            return false;
        }
    }
}
```

The preceding code snippet parses and sets each cell value (firstname, lastname, and so forth) and reads each line from the .csv file like this:

```
CsvEntry entry = new CsvEntry();
while ((line = reader.readLine()) != null)
       {
           if (entry.parse(line, lineNumber))
           {

...
}
}
```

7. After parsing, let's prepare a composite type for each column value. Since we have uuid, int, and text as data types defined for the primary key, first name, last name, and age, we need to define them as follows:

```
ByteBuffer uuid = UUIDType.instance.decompose(entry.key);
usersWriter.newRow(uuid);

List types = newArrayList();
types.add(UTF8Type.instance);

CompositeType compositeType = CompositeType.getInstance(types);

List numericTypes = newArrayList();
numericTypes.add(IntegerType.instance);

CompositeType numericType = CompositeType.getInstance(numericTypes);
```

8. Next, let's assign values:

```
long timestamp = System.currentTimeMillis() * 1000; usersWriter.
addColumn(compositeType.decompose("firstname"),
                    compositeType.decompose(bytes(entry.firstname)), timestamp);
usersWriter.addColumn(compositeType.decompose("lastname"),
compositeType.decompose(bytes(entry.lastname)), timestamp);
             usersWriter.addColumn(compositeType.decompose("password"),
compositeType.decompose(bytes(entry.password)), timestamp);

usersWriter.addColumn(compositeType.decompose("age"),
numericType.decompose(bytes(entry.age)),timestamp);

usersWriter.addColumn(compositeType.decompose("email"),
compositeType.decompose(bytes(entry.email)),timestamp);
```

9. Now, to assign the addresses collection, we need to construct and assign with data type as Set:

```
SetType<String> addressType = newSetType<String>(UTF8Type.instance);

// column name
ByteBuffer byteBuffer1 = compositeType.decompose("addresses");

// set value
ByteBuffer byteBuffer2 = compositeType.decompose("noida");
```

10. We need to create a bytebuffer instance and add a column:

```
ByteBuffer bb = ByteBuffer.allocate(byteBuffer2.capacity()+byteBuffer1.capacity());
bb.put(byteBuffer1);
bb.put(byteBuffer2);
bb.flip();

usersWriter.addColumn(bb,addressType.decompose(new HashSet<String>()),timestamp);
```

11. Next, we can save it:

```
usersWriter.close();
```

The preceding command generates the .db files under the data/Cql3Demo/Users folder.

12. Now, we need to load the generated .db files using sstableloader:

```
$CASSANDRA_HOME/bin/sstableloader -d localhost data/Cql3Demo/Users/
```

13. We can validate the output by issuing the select command using the cql shell (see Figure 10-3).

```
select * from "Users";
```

```
                  user_id   | addresses | age | email                        | firstname | lastname | password
--------------------------------------+-----------+-----+------------------------------+-----------+----------+----------
5bd8c586-ae44-11e0-97b8-0026b0ea8cd0 |  {noida}  |  22 | vivek.mishra@nomail.com      |     vivek |   mishra |    4Jc2s
4bd8cb58-ae44-12e0-a2b8-0026b0ed9cd1 |  {noida}  |  12 |         bigdata@nomail.com   |    apress |  bigdata |    s!a0ml
1ce7cb58-ae44-12e0-a2b8-0026b0ad21ab |  {noida}  |  12 |    brian.neil@nomail.com     |     Brian |     Neil |    s)3B3
```

Figure 10-3. *The Users table with three rows*

14. We can also alter the **Users** table and add a column with the data type of map:

```
alter table "Users" add mapCol  map<text,text>;
```

15.   We just need to add the `composite` type for map and add map as a column to the preceding implementation like this:

```
// map column name
byteBuffer1 = compositeType.decompose("mapcol");
// map column key
byteBuffer2 = compositeType.decompose("noida");

bb =ByteBuffer.allocate(byteBuffer2.capacity()+byteBuffer1.capacity());

bb.put(byteBuffer1);
bb.put(byteBuffer2);
bb.flip();

// map value
usersWriter.addColumn(bb,compositeType.decompose(bytes("vivek")),timestamp);
```

16.   Issuing a `select` query over **Users** will produce the output shown in Figure 10-4.

```
              user_id          | addresses | age | email                  | firstname | lastname |   mapcol       | password
--------------------------------+-----------+-----+------------------------+-----------+----------+----------------+----------
5bd8c586-ae44-11e0-97b8-0026b0ea8cd0 |  {noida} |  22 | vivek.mishra@nomail.com |     vivek |   mishra | {noida: vivek} |    4Jc2s
4bd8cb58-ae44-12e0-a2b8-0026b0ed9cd1 |  {noida} |  12 |        bigdata@nomail.com |    apress |  bigdata | {noida: vivek} |   s!a0ml
1ce7cb58-ae44-12e0-a2b8-0026b0ad21ab |  {noida} |  12 |   brian.neil@nomail.com |     Brian |     Neil | {noida: vivek} |   s)3B3
```

Figure 10-4. *Output with a map column in the Users table*

So that's how we can bulk load `.db` files directly on to disk in sstable format.

Summary

Just to summarize, in this chapter we discussed the `nodetool` utility's usage and various operations for ring and schema management. We also discussed how to import and export data in JSON format. Finally we discussed loading bulkloading CQL3 format data (including collection).

In the next and final chapter, we will be summarizing our discussion and talking about Cassandra version upgrades and other popular discussions in various forums.

CHAPTER 11

■ ■ ■

Upgrading Cassandra and Troubleshooting

So far, topics discussed in this book have included configuration, batch analytics, data modeling techniques, performance tuning, and various utilities. One important topic yet to be discussed is Cassandra version upgrades. The past one and half years have seen quick development and multiple Cassandra releases. Some of the changes introduced in recent releases have backward compatibility issues, which makes Cassandra version upgrading an important topic that needs to be addressed. The most recent release, at the time of this writing, is version 2.1, which includes a couple important new features we have not covered elsewhere.

We have discussed various troubleshooting tips in different places in the book. In this chapter, we will summarize several key ones that address issues you may well encounter.

The topics in this chapter include:

- An overview of the Cassandra 2.1 release
- Cassandra version upgrade
- Troubleshooting tips
- The road ahead with Cassandra

Let's start by discussing the new features introduced with the recently released Cassandra 2.1.

Cassandra 2.1

The recently released Cassandra 2.1 enables a few important features along with many performance-related fixes. In this section, we will discuss three important features supported by Cassandra 2.1:

- User-defined types
- Frozen types
- Indexing on collection attributes

■ **Note** For a complete overview of what's new in Cassandra 2.1, see the following documentation page: www.datastax.com/documentation/cassandra/2.1/cassandra/features2.html.

User-Defined Types

User-defined types allow you to combine multiple fields of information in a single table. For example, you can have a single field with information from multiple fields using the CQL3 shell like this:

```
create type if not exists twitter_keyspace.user_metadata(fname,lname)
```

Here user_metadata is a user-defined type which is a wrapper on multiple columns. With user-defined types, we can simplify data modeling schema by reducing multiple tables to fewer tables.

Frozen Types

Using the frozen keyword, we can define a particular user-defined type as follows:

```
CREATE TABLE twitter_keyspace.users (
  user_id timeuuid PRIMARY KEY,
  name frozen <user_metadata>,
);
```

Here, name is a frozen type of user_metadata for the users table in the keyspace **twitter_keyspace**. With a column field defined with the frozen keyword, Cassandra serializes multiple components in single value. For example, with user_metadata, fname and lname will be serialized as a single value. Once a user-defined data type is frozen with the frozen keyword, we cannot update components of the user-defined type.

Indexing on Collection Attributes

In Chapter 2 we discussed data modeling concepts and working with collections. Also under the "Secondary Indexes" section in Chapter 2, we discussed that indexes over Cassandra collection attributes were not supported. With Cassandra 2.1, we now can create indexes over Cassandra collection attributes like this:

```
create table users(user_id text PRIMARY KEY,fullname text,email text,password text, followers
map<text, text>);
insert into users(user_id,email,password,fullname,followers) values ('imvivek','imvivek@xxx.com',
'password','vivekm',{'mkundera':'milan kundera','guest': 'guestuser'});

create index followers_idx on users(followers);
```

Here, users is a table and followers is a map containing user_id as its key and name as the value. The create index command will enable an index over the values of the followers map (e.g., full name). We can also enable indexes over followers map keys (e.g., user_id) like this:

```
create index followers_idx_keys on users(KEYS(followers));
```

We can fetch records from followers using indexes as follows:

```
SELECT email, followers FROM users  WHERE followers CONTAINS 'milan kundera';
```

This command will return all followers having the full name "milan kundera".

One more change worth discussing is that Cassandra 2.1 doesn't support the sstable format of versions earlier than Cassandra 2.x, which leads to the important question of how to handle version upgrades! In the next section, we will discuss how to handle version upgrades with Cassandra.

Upgrading Cassandra Versions

In the last year and a half, Cassandra has seen the rapid releases of multiple versions of Cassandra, including 1.2.x, 2.0.x, and 2.1.x. The reasons for the parallel development on multiple versions were maintenance, priority bug fixes, and rolling out version 2.0. Version 2.0 came with many new feature sets, which we discussed in Chapter 2.

Applications using Cassandra as a database also required upgrades to their newest releases. Ideally it should be a smooth ride, but major version changes came with format changes in sstables and brought in version restrictions.

The volume of releases since last summer is displayed in Figure 11-1, which shows the many Cassandra releases since August 2013 and that parallel development is required on the Cassandra side for maintenance and feature releases. Maintenance releases can also be termed as minor releases, such as critical bug fixes, whereas major releases focus primarily on new feature support and performance fixes.

3 days ago	**cassandra-2.1.2** … ○ cdf8ed9 🗎 zip 🗎 tar.gz	on Jul 3	**cassandra-1.2.18** … ○ 5a658be 🗎 zip 🗎 tar.gz	on Feb 3	**cassandra-1.2.14** … ○ 6a93144 🗎 zip 🗎 tar.gz
20 days ago	**cassandra-2.0.11** … ○ 02b83d9 🗎 zip 🗎 tar.gz	on Jun 30	**cassandra-2.0.9** … ○ 5b878ce 🗎 zip 🗎 tar.gz	on Dec 30, 2013	**cassandra-2.0.4** … ○ d56f8f2 🗎 zip 🗎 tar.gz
20 days ago	**cassandra-2.1.1** … ○ 3261d5e 🗎 zip 🗎 tar.gz	on Jun 30	**cassandra-1.2.17** … ○ 87c4efe 🗎 zip 🗎 tar.gz	on Dec 19, 2013	**cassandra-1.2.13** … ○ 1b4c9b4 🗎 zip 🗎 tar.gz
on Sep 18	**cassandra-1.2.19** … ○ 2d29ebd 🗎 zip 🗎 tar.gz	on Jun 26	**cassandra-2.1.0-rc2** … ○ e2befe2 🗎 zip 🗎 tar.gz	on Nov 25, 2013	**cassandra-2.0.3** … ○ 3c9760b 🗎 zip 🗎 tar.gz
on Sep 18	**cassandra-2.1.0-deb** … ○ ee6c1e8 🗎 zip 🗎 tar.gz	on Jun 2	**cassandra-2.1.0-rc1** … ○ c108ce5 🗎 zip 🗎 tar.gz	on Nov 25, 2013	**cassandra-1.2.12** … ○ 026865f 🗎 zip 🗎 tar.gz
on Sep 16	**cassandra-2.1.0** … ○ c6a2c65 🗎 zip 🗎 tar.gz	on May 29	**cassandra-2.0.8** … ○ 484d281 🗎 zip 🗎 tar.gz	on Oct 28, 2013	**cassandra-2.0.2** … ○ 22e67be 🗎 zip 🗎 tar.gz
on Sep 3	**cassandra-2.1.0-rc7** … ○ 02be0de 🗎 zip 🗎 tar.gz	on May 5	**cassandra-2.1.0-beta2** … ○ 48727b4 🗎 zip 🗎 tar.gz	on Oct 22, 2013	**cassandra-1.2.11** … ○ 9424449 🗎 zip 🗎 tar.gz
on Aug 25	**cassandra-2.0.10** … ○ e28e7bf 🗎 zip 🗎 tar.gz	on Apr 18	**cassandra-2.0.7** … ○ 7dbbe92 🗎 zip 🗎 tar.gz	on Sep 23, 2013	**cassandra-2.0.1** … ○ eb96db6 🗎 zip 🗎 tar.gz
on Aug 19	**cassandra-2.1.0-rc6** … ○ d087317 🗎 zip 🗎 tar.gz	on Mar 31	**cassandra-1.2.16** … ○ 05fcfa2 🗎 zip 🗎 tar.gz	on Sep 22, 2013	**cassandra-1.2.10** … ○ 9375363 🗎 zip 🗎 tar.gz
on Aug 4	**cassandra-2.1.0-rc5** … ○ cfb335e 🗎 zip 🗎 tar.gz	on Mar 10	**cassandra-2.0.6** … ○ 656edc5 🗎 zip 🗎 tar.gz	on Sep 3, 2013	**cassandra-2.0.0** … ○ 03045ca 🗎 zip 🗎 tar.gz
on Jul 19	**cassandra-2.1.0-rc4** … ○ d872e2c 🗎 zip 🗎 tar.gz	on Feb 20	**cassandra-2.1.0-beta1** … ○ 73dcdbd 🗎 zip 🗎 tar.gz	on Aug 30, 2013	**cassandra-1.2.9** … ○ 6164d01 🗎 zip 🗎 tar.gz
on Jul 10	**cassandra-2.1.0-rc3** … ○ 0bc9841 🗎 zip 🗎 tar.gz	on Feb 6	**cassandra-1.2.15** … ○ 178e086 🗎 zip 🗎 tar.gz	on Aug 20, 2013	**cassandra-2.0.0-rc2** … ○ 3e516a3 🗎 zip 🗎 tar.gz
on Jul 3	**cassandra-1.2.18** … ○ 5a658be 🗎 zip 🗎 tar.gz	on Feb 6	**cassandra-2.0.5** … ○ 29670eb 🗎 zip 🗎 tar.gz	on Aug 8, 2013	**cassandra-2.0.0-rc1** … ○ e8ae672 🗎 zip 🗎 tar.gz

Figure 11-1. *Cassandra release chart since August 2013*

Backward Compatibility

A Cassandra release is backward compatible if it works fine with input and the data model obtained with previous releases. Ideally, each future release should be backward compatible to avoid administrative troubleshooting. Significant changes have been made to Cassandra's storage architecture, however, with rapid releases of new feature support and performance enhancements. It is extremely important to understand the compatibility issues across the various Cassandra releases and in some cases the sequence you have to follow to avoid problems.

In general, Cassandra requires you to migrate through major and minor releases in sequence. The reason for such sequential upgrades is that every release would educate you on compatibility issues and its version upgrade process. For instance, Cassandra 2.0 releases are not compatible with versions older than the 1.2.9 release. Similarly, upgrading from a 1.2.x Cassandra release to 2.1 requires doing a rolling restart to version 2.0.7 followed by version 2.1. Without the rolling restart, you should see the following error in the Cassandra server log:

```
java.lang.RuntimeException: Can't open incompatible SSTable! Current version jb, found file:
/var/lib/cassandra/data/system/schema_columnfamilies/system-schema_columnfamilies-ib-5
    at org.apache.cassandra.db.ColumnFamilyStore.createColumnFamilyStore(ColumnFamilyStore.java:410)
    at org.apache.cassandra.db.ColumnFamilyStore.createColumnFamilyStore(ColumnFamilyStore.java:387)
    at org.apache.cassandra.db.Keyspace.initCf(Keyspace.java:309)
    at org.apache.cassandra.db.Keyspace.<init>(Keyspace.java:266)
    at org.apache.cassandra.db.Keyspace.open(Keyspace.java:110)
    at org.apache.cassandra.db.Keyspace.open(Keyspace.java:88)
    at org.apache.cassandra.db.SystemKeyspace.checkHealth(SystemKeyspace.java:499)
    at org.apache.cassandra.service.CassandraDaemon.setup(CassandraDaemon.java:228)
```

Hence it's recommended to follow release notes to properly address version upgrade-related issues. As discussed above, to solve such version upgrade issues, we need to perform a rolling restart. A rolling restart is one that doesn't bring down the cluster and perform a version upgrade on each node, but rather one that performs the upgrade on nodes with zero downtime. Nodes get upgraded and restarted one at a time so that data availability can still be assured. This means a node with Cassandra 1.1.x version, for example, would require first an upgrade to Cassandra version 1.2.9 and then a version 2.0 upgrade like the one mentioned previously: that is, the sequence would be version 1.1.x to 1.2.9 to 2.0.7 and finally to 2.1.

A few of the configuration changes in version 2.0 release are

- The property `index_interval` has been moved to the table level and is no longer available in `cassandra.yaml`.

- Virtual nodes (e.g., `num_tokens`) are enabled by default with 2.0 and later versions whereas with earlier versions it was disabled.

- Java version 7 must be installed for 2.0 and later releases.

Now that we have discussed backward compatibility, let's see how to perform a version upgrade on Cassandra nodes.

Performing an Upgrade with a Rolling Restart

Using the scenario mentioned in the preceding section, let's walk through the steps for performing the sequential upgrade:

1. Before we start the rolling upgrade, we should take care of the above-mentioned changes and take a backup of all configurations belonging to the previous Cassandra version and data. For data backup we can create a snapshot as follows:

   ```
   vivek@vivek-Vostro-3560:~$CASSANDRA_HOME$ bin/nodetool -h localhost snapshot
   twitter_keyspace
   ```

In the preceding command, please replace $CASSANDRA_HOME$ with the currently installed Cassandra version (e.g., $apache-cassandra-1.1.6$).

In the preceding command, we are creating a snapshot for keyspace **twitter_keyspace**.

2. Please make sure to remove all dead nodes from the running Cassandra cluster before downloading the next version, 1.2.9 in this case.

3. Then run the following command:

```
nodetool upgradesstables
```

This will upgrade the existing sstables (with version 1.1.x) to a format compatible with 1.2.9.

4. Then, in sequence, follow the preceding steps for upgrading from 1.2.9 to 2.0.7, and then finally, perform the same for the upgrade from version 2.0.7 to 2.1.

For a multi-node Cassandra cluster we need to perform the same steps over all the nodes. If version upgrades fail on any node, we can try copying schema from another node and replicate the data from other live nodes.

Troubleshooting Cassandra

Troubleshooting is the process of diagnosing and fixing problems. In this section, we will troubleshoot a few common problems—such as having too many open files and running out of memory—that you might encounter and have to troubleshoot.

Too Many Open Files

If you get an error like

java.net.SocketException: Too many open files
at java.net.Socket.createImpl(Socket.java:447)
at java.net.Socket.getImpl(Socket.java:510)
at java.net.Socket.setSoLinger(Socket.java:984)

it means you are hitting the limit of the maximum allowed open files. The default value is 1024. With Cassandra, it is highly possible to exceed this limit. We can increase this limit by modifying the **/etc/security/limits.conf** file like this:

```
root soft nofile 65535
```

Here 65535 is the number of file descriptors open. This should solve the issue. Additionally, we can monitor the number of file descriptors by monitoring running Java processes to validate whether it is a case of memory leak or not! Using the lsof command, we can list all open files. For example, we can list all open files for a process ID (PID) as follows:

lsof -p 13241

```
COMMAND    PID       USER    FD     TYPE    DEVICE    SIZE/OFF    NODE NAME
init       13241     root    cwd    DIR     225,0     4096        1 /
init       13241     root    rtd    DIR     225,0     4096        1 /
```

Stack Size Limit

The stack size limits the number of threads we can have in the Java Virtual Machine (JVM). If you are seeing an error like the following one:

```
The stack size specified is too small, Specify at least 228k
Error: Could not create the Java Virtual Machine.
Error: A fatal exception has occurred. Program will exit.
```

it means stack size is set too low. The default assigned value is 180K, and as recommended by Cassandra, we can set it to 228K by modifying the Cassandra-env.sh file (available under the conf folder) like this:

```
JVM_OPTS="$JVM_OPTS -Xss228k"
```

Out of Memory Errors

With the massive volume of data processed with Cassandra, sometimes nodes may die with "out of memory" errors. If you encounter this situation, first check the row cache and memtable sizes and validate whether they are set at values that are too large. Here, "too large" simply means they can't fit well in the available memory. If these values are well below available memory limits, then we should investigate the heap dump and server and application logs for further information on errors. With Cassandra 1.2.9 onwards, many background processes have been moved from on-heap to off-heap processing to avoid garbage collection activities. So you can run out of memory off-heap or on-heap. For more discussion about off-heap and on-heap, please refer the "CPU and Memory Utilization" and "Off-Heap vs. On-Heap" sections in Chapter 8. Additionally, you can check client-side code in case the application is reading something like entire rows or large columns.

Too Much Garbage Collection Activity

If garbage collection (GC) takes longer than a few seconds, check the system.log file for a message like one of the following:

```
INFO [ScheduledTasks:1] 2014-01-29 02:41:16,579 GCInspector.java (line 116) GC for
ConcurrentMarkSweep: 341 ms for 1 collections, 8001582816 used; max is 8126464000

INFO [ScheduledTasks:1] 2014-01-29 02:41:29,135 GCInspector.java (line 116) GC for
ConcurrentMarkSweep: 350 ms for 1 collections, 8027555576 used; max is 8126464000
```

Long garbage collection pauses can create a stop-the-world scenario that brings data nodes to a freeze state for some time. In such scenarios, it is highly possible that a portion of the JVM that is not in use would be swapped out. It is recommended to keep this off using

```
swapoff -all
```

Running this command over a Linux terminal will disable swapping for all devices.

These are a few common problems you are likely to encounter and their solutions. Next let's discuss the road ahead with Cassandra.

Road Ahead with Cassandra

Apache Cassandra is still under active development, and an extremely active community keeps the momentum going. We have already discussed the multiple tools and APIs available for Apache Cassandra development. Figure 11-2 shows a chart of upcoming Cassandra releases at the time of this writing (taken from `https://issues.apache.org/jira/browse/CASSANDRA/fixforversion/12324945/?selectedTab=com.atlassian.jira.jira-projects-plugin:version-summary-panel`).

Versions: Unreleased

Name	Release date
📦 2.0.12	
📦 3.0	
📦 3.1	
📦 2.1.3	

Figure 11-2. Upcoming Cassandra releases (as of November 2014)

There is absolutely no doubt that Cassandra has been widely accepted as a preferred columnar database for building scalable big data applications. Companies such as eBay, Netflix, and Facebook are using Cassandra for multiple purposes. Developers from these organizations have been regularly contributing and introducing their Cassandra-based tools, which bodes well for the future.

Furthermore, there are high-level clients available in almost all language options. Table 11-1 shows a table listing popular high-level Cassandra client APIs.

Table 11-1. Cassandra High-Level Clients

Language	Clients
Java	DataStax Java driver, Astyanax, Hector, Kundera
Python	DataStax Python driver, pycassa
.NET	DataStax C# driver, CqlSharp
C++	libcql
Scala	scqla
Perl	perlcassa
PHP	Cassandra PDO driver, phpcasa

With multiple APIs in different languages, an active community, and planned releases, Cassandra definitely has a long road ahead. All-new sets of interesting features are expected in upcoming Cassandra releases. Again, Cassandra is a popular NoSQL database that is ideal for dealing for handling big data; the ecosystem is rich in client APIs for Cassandra; active development that has been going on for several years continues; and companies like Netflix, eBay, and Facebook are using and promoting it. With many new commercial tools and products coming in this space, it's a database to learn and use!

Summary

With this we come to the end of this chapter and the book. From Chapter 1 to Chapter 11, we have discussed different Cassandra features and their implementations in the form of sample exercises. In this chapter, we looked at the process of upgrading to new versions of Cassandra and how to navigate the complexities of so many recent releases. We also considered several common and important problems you likely will have to troubleshoot at some point, and we looked at the road ahead for Cassandra.

Key topics covered earlier in the book included batch analytics and available tools (Chapters 5 and 6), graph-based implementations over Cassandra (Chapter 7), and performance tuning and various performance-related do's and don'ts (Chapter 8).

With this book, I tried to keep a balance between sample recipes and theory. I hope you have enjoyed it and learned things as you expected. Thanks!

References

The following list includes a few of the references I used while writing this book. They can be useful for keeping an eye on Cassandra's future development:

- http://cassandra.apache.org/

- www.datastax.com/

- http://wiki.apache.org/cassandra/

- http://planetcassandra.org/blog/

- For a general discussion mailing list, subscribe to the user mailing list at user-subscribe@cassandra.apache.org.

- For development-related discussion, you can subscribe to dev-subscribe@cassandra.apache.org.

- To discuss client API-related topics, subscribe to client-dev-subscribe@cassandra.apache.org.

Index

■ E, F

■ G

Get the eBook for only $10!

Now you can take the weightless companion with you anywhere, anytime. Your purchase of this book entitles you to 3 electronic versions for only $10.

This Apress title will prove so indispensible that you'll want to carry it with you everywhere, which is why we are offering the eBook in 3 formats for only $10 if you have already purchased the print book.

Convenient and fully searchable, the PDF version enables you to easily find and copy code—or perform examples by quickly toggling between instructions and applications. The MOBI format is ideal for your Kindle, while the ePUB can be utilized on a variety of mobile devices.

Go to www.apress.com/promo/tendollars to purchase your companion eBook.